MILTON D. HEIFETZ &

A WALK THROUGH THE

SOUTHERN SKY

A Guide to Stars and Constellations and their Legends

CAMBRIDGE
UNIVERSITY PRESS

PUBLISHED BY THE PRESS SYNDICATE OF THE UNIVERSITY OF CAMBRIDGE
The Pitt Building, Trumpington Street, Cambridge, United Kingdom

CAMBRIDGE UNIVERSITY PRESS
The Edinburgh Building, Cambridge CB2 2RU, UK   http://www.cup.cam.ac.uk
40 West 20th Street, New York, NY 10011-4211, USA   http://www.cup.org
10 Stamford Road, Oakleigh, Melbourne 3166, Australia http://www.cup.edu.au
Ruiz de Alarcón 13, 28014 Madrid, Spain

First published 2000

Printed in Australia by Brown Prior Anderson

Typeset in New Century Schoolbook 10/14pt *System* QuarkXpress® [SE]

A catalogue record for this book is available from the British Library

Library of Congress Cataloguing in Publication data
Heifetz, Milton D., 1921–
A walk through the southern sky : a guide to stars and constellations and their legends /
Milton D. Heifetz & Wil Tirion.
     p.   cm.
ISBN 0 521 66514 0 (pb)
1. Astronomy – Southern Hemisphere – Observers' manuals.   2. Constellations.   I. Tirion,
Wil.   II. Title.
QB63.H376 1999
523.8′02′16–dc21   99-0499884

ISBN 0 521 66514 0 paperback

Contents

Contents

Contents

Contents

Acknowledgements

I wish to acknowledge the courtesy of Larry Schindler and Ron Dantowitz for use of the Hayden Planetarium of Boston. I also wish to express my gratitude to Robert Kimberk and Freeman Deutsch of the Harvard University Center of Astrophysics for their generous computer assistance and to Robert S. Strobie, Brian Warner and John Menzie of the South African Astronomical Observatory in Sutherland for their courtesy during my study of the Southern Constellations. I also wish to acknowledge the excellent advice I received from Dan Ben-Amos, Yosef Dan, Jarita Holbrook, Jan Knappert, Edwin Krupp, Harold Scheub, Gregory Schremp, and Gary Urton regarding sources for the legends of Africa, South America and the Pacific Islands, and especially Suzanne Blair, who allowed me complete access to her voluminous research on the cosmological folklore of the natives of Africa.

Any errors are strictly my own and the artistic license I may have taken in relation to the legends is also my own responsibility.

Introduction

This book is written for those who look at the stars with wonderment and would like to feel more at home with them, to go for a friendly walk with them.

In order to walk through the heavens and to know where you are, you must recognize what your eye sees. To know the names of stars and constellations is to form a friendship with our heavenly neighbors.

As we walk among the constellations, you will feel the immensity and quiet peace of the night sky. Do not ignore the legends about the constellations in Part 3 of the book. These legends will lend greater feeling to your vision of the world above. Friendship with the stars will deepen as we sense the thoughts and dreams of people who imagined people and animals living among the constellations.

Our walk will take us to the brightest stars in the sky. When we become familiar with them they will lead us to the dim stars.

It is not enough simply to find a constellation. Try to see relationships between constellations. This is best done if you know different pathways to the constellations.

From the time of early humans, people have looked at the stars to help them navigate across seas and deserts, know when to plant and to harvest, establish their legends, mark the change of seasons and even align their temples of worship. To aid in recognizing specific stars they placed the brighter ones into star group patterns we now call **constellations**. Constellations were recorded over 5000 years ago and lists of such patterns were written 2400 years ago by the Greek astronomer Eudoxus who studied under Plato. Ptolemy, who lived 2100 years ago, compiled a list of 48 constellations which has remained relatively standard to this day. Later, Johann Bayer (1572–1625), Johannes Hevelius (1611–1687) and Nicolas de Lacaille (1713–1762) added more constellations to the list. Professional astronomers now officially recognize 88 constellations which they regard simply as areas of the sky, not as star 'pictures' or patterns. These patterns have never been made 'official', so you should feel free to make any constellation design you wish.

Before we begin our walk through the heavens, we should understand two concepts: how to measure distances in the sky, and the brightness of the stars. After this is done, follow the instructions on how to use the atlas to best advantage.

In Part 2, 'A walk through the heavens', the design or picture of a group of stars to form a constellation image will usually, but not always, contain stars which are bright enough to be seen easily. Most of the constellation patterns are well recognized images, but some are new.

For convenience, each star in each constellation will be numbered and some will be named so that we can more easily identify specific stars to help us walk around the sky. We will follow several paths to a constellation. By doing this you will have a better sense of star relationships.

Since I have been disturbed with the violence that is part of the commonly used legends associated with the constellations, I have taken the liberty of modifying and abridging them. Legends have been and will continue to be modified with each generation.

This book applies to people living in the Southern Hemisphere, but it is also of value to those living slightly north as well as south of the equator.

Relax and enjoy yourself as you travel across the sky.

Part 1

Measuring distances in the sky

How do we measure the size of Scorpius or the distance between two stars? We cannot measure these distances in inches or millimeters, which are linear measurements (measurements along a line). Instead, we must use a measuring system using angles to determine how far apart one star or constellation may be from another.

To do this in a practical way without fancy instruments we use our eye as the corner of the angle and part of our hand to hide the sky between the stars or constellations of interest. The further apart the stars are, the more of our hand we need to use to cover the space between them. Look at Fig. 1.

Fig. 1

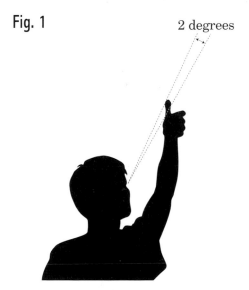

2 degrees

With your arm outstretched, your hand will help you determine angular distances. Extend your arm out in front of you and hold your thumb upright. It is now hiding part of what is in front of your vision. The amount of view that is hidden behind your thumb will depend upon how long your arm is and how thick your thumb may be. The shorter the arm, or the thicker the thumb, the more of your view will be hidden.

So our hand becomes an excellent device for measuring distances in degrees in the sky. Different parts of your hand can be used to measure different angles. Look at Fig. 2.

Fig. 2

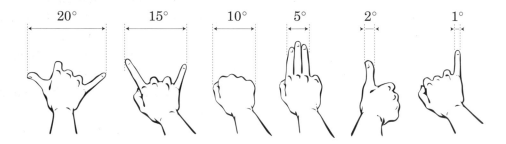

| 20° | 15° | 10° | 5° | 2° | 1° |

The tip of your small finger will cover approximately 1 degree of sky. In your room look at the door knob or light switch across the room. Your finger can cover it. Now look at a building across the street. The same finger will cover a large part of the building. Now look at the Moon. The same finger can cover the Moon. How can this be since one is so much larger than the other? Look at Fig. 3.

Fig. 3

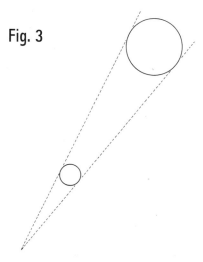

Although the Moon is so much larger than the building across the street, it can actually be hidden by a narrow object like a finger. The diameter of the Moon, when measured this way, is seen to be only about ½ degree wide. The farther away an object is, the smaller the angle needed to hide it from sight. The Moon looks much bigger than a star because it is so much closer to us.

Distances to the stars

We measure the distance between a star and the Earth, not in miles or kilometers, but in **light years** by using the speed of light. It is important to remember that a light year is a distance, it is not a measure of time. The distance light travels in one year is a light year. Light travels 186 000 miles per second (299 000 kilometers per second), which is 680 760 000 miles per hour (1 096 000 000 kilometers per hour). Therefore, a light year is a distance of almost 6 000 000 000 000 (6 trillion) miles, or 9.6 trillion kilometers.

It takes more than one second for light from the Moon to reach the Earth and more than 8 minutes for light from the Sun to reach Earth. Compare this with the 4.3 years that it takes for the light from the nearest star, Rigil of Centaurus, to reach the Earth. Deneb in the Northern Cross is over 1000 light years away. That means the light we now see left the star over 1000 years ago. It is therefore possible that the star may not even be there any more.

The brightness of stars

Some stars appear much brighter than others. This does not necessarily mean that the bright star is bigger or giving off more light than the dimmer star. The **apparent brightness** (how bright it seems to us) depends upon three things:

(1) how big it is; (2) how far away it is from Earth; and (3) how much light it actually emanates per diameter of the star. The brightest star to us is our Sun, but it is only an average size star. It seems the brightest because it is the nearest star to us on Earth.

The star Sirius in the constellation of Canis Major appears considerably brighter than Rigel in Orion. However, Rigel is actually thousands of times brighter than Sirius. It appears fainter because it is over a thousand light years away, while Sirius is only 8½ light years from us.

We measure the brightness of the stars as seen with the naked eye on a scale called the **magnitude** scale. Hipparchus, a Greek astronomer, rated the importance of stars by their brightness and used the word magnitude to describe their relative brightness. Magnitude means bigness. In ancient times they may have assumed that the brighter star is a bigger star. A very bright star would have a magnitude of 1 or less and a very faint star a magnitude of 6. The smaller the number, the brighter the star. A very powerful telescope can see very faint stars beyond magnitude 20. You may be able to see stars with a magnitude of 6 to 7 with your naked eye under very clear, moonless skies. The very brightest planets have a magnitude of -1 to -4. Unfortunately, light pollution from home and street lamps may prevent you from seeing as many stars as you could if your surroundings were in total darkness. Remember, magnitude is a measure of star brightness, not how much light the star actually produces, nor how big it is.

Although there are billions of stars, we can only see approximately 2500 stars with our naked eye at one time under the best of conditions. Read how to test your vision in Part 4.

The Milky Way

The space around us seems to be endless. It is a space occupied by billions upon billions of galaxies, each of which is composed of billions and billions of stars, of which our Sun is an average-sized example. The faint band of stars that arches across the sky was called the Milky Way by the early Greeks. It is our view of the galaxy in which we live from within one of our galaxy's spiral arms. The location of our Sun and Earth in that spiral arm is approximately 30 000 light years from the center of our galaxy.

What we see with our naked eye is confined to our own galaxy. However, with good eyesight, and if the night is dark enough, you may see a neighboring galaxy as a faint blur in the constellation Andromeda, or the Small and Large Magellanic Clouds in the region of the constellation Hydrus.

Although our galaxy is whirling in space at tremendous speed it still takes 225 million years to complete one revolution. That time period is called the Galactic Year.

Imagine yourself sitting near the end of a spiral arm of our galaxy. If you look straight up or down you will see neighboring stars in our spiral arm of our galaxy but when you look toward Sagittarius you are looking along the flat side of the spiral arm toward the wider and more dense center bulge of our galaxy. We cannot see the spiral arm opposite us because it is hidden by the billions of stars in the center of the galaxy.

As you look along the Milky Way you will notice that some areas appear to have dark holes or slots in it. These are not empty spaces, but rather dark masses of dust, star debris and gases that hide the stars behind them. There is a very definite dark slit in the area of Cygnus, sometimes called the Cygnus Rift, or the Northern Coalsack, and a similar dark patch in area of the Southern Cross, called the Southern Coalsack.

Fig. 4 – Side view of a galaxy

Fig. 5 – Top view of our galaxy, showing the position of our Sun

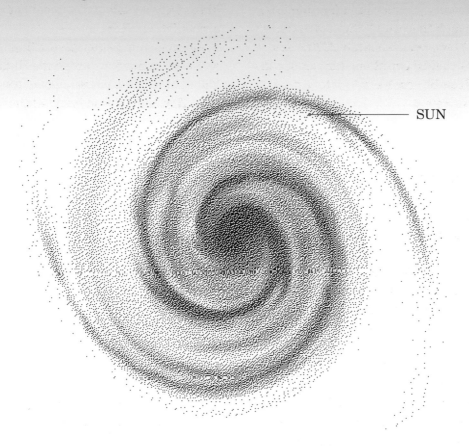

SUN

Life in the heavens

There are nine known planets orbiting our Sun. There are over 100 000 000 000 (100 billion) suns (stars) in our galaxy. There are billions of galaxies. Just imagine how many planets there must be in our galaxy alone.

In 1995, three planets were almost certainly identified in distant stars. Two of them appear to be as far from their Sun as Earth is from our Sun. This suggests that at least as far as distance to the life-giving source of a Sun is concerned they are not unlike Earth. It is also important to realize that the basic elements necessary for life as we know it, carbon, hydrogen, oxygen and nitrogen, exist throughout the heavens, and that amino acids have been found in meteorites. Given the proper environmental conditions these molecules may join to form the proteins and RNA of living cells, which can then replicate themselves. Such action would signify life.

It is almost impossible to assume that we are the only planet with life on it. There are probably massive colonies of microorganisms living in the deep rock structures of planets. The question is not whether there are any living organisms among the stars, but rather, what kind of life is there and are they trying to contact us?

Instructions for use of the atlas

Begin your walk through the sky by first determining which constellations are visible overhead during the month of your walk. Look at the following four star charts which give an overview of the constellations visible during each of the four seasons. Hold the chart in front of you like reading a book. Do not place it overhead. You will notice north or south marked at the bottom of the diagram. Face north or south and compare the lower half of the chart with the stars in the sky. Then face east or west and turn the star chart around so that the direction you are facing is at the bottom of the chart. When you have determined which constellation you wish to see, turn to the index, which will identify the diagrams concerning that constellation. There is a date at the top of each diagram. Even though you may be looking at the sky on a different date, the positions of the constellations will not change in relationship to each other. Those relationships are constant, but, due to the Earth's rotation, you may have to turn the diagram in order for the constellations to appear as they do at your time and date. If you are not familiar with the constellations then start your walk with Crux on page 15. If you are familiar with some of the star groups, then, depending upon the season, look for a specific constellation by using the index to find the diagram dealing with that constellation's relationship to other constellations. If you are slightly familiar with some constellations these four charts can help you find others.

Spring Stars

September–October 10–12 pm
November–December 7–9 pm

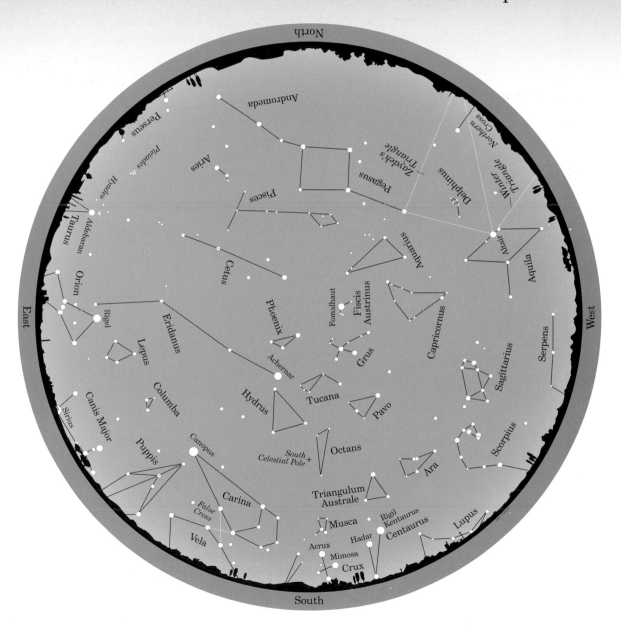

Summer Stars

December–January 10–12 pm
February–March 7–9 pm

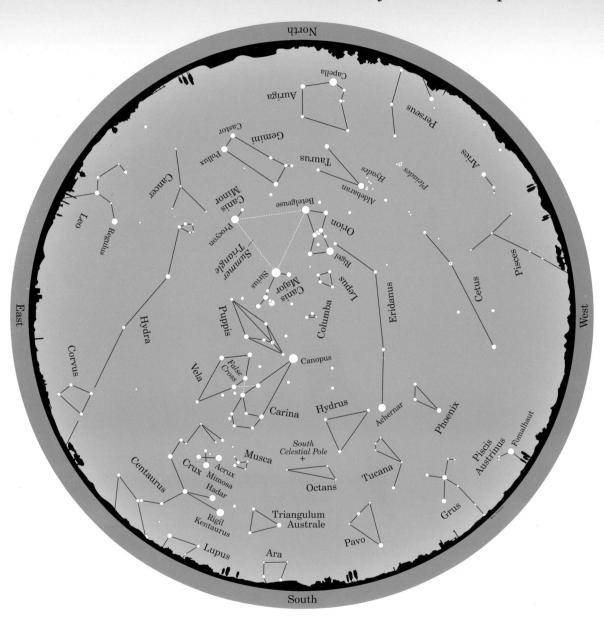

Autumn Stars

March–April 10–12 pm
May–June 7–9 pm

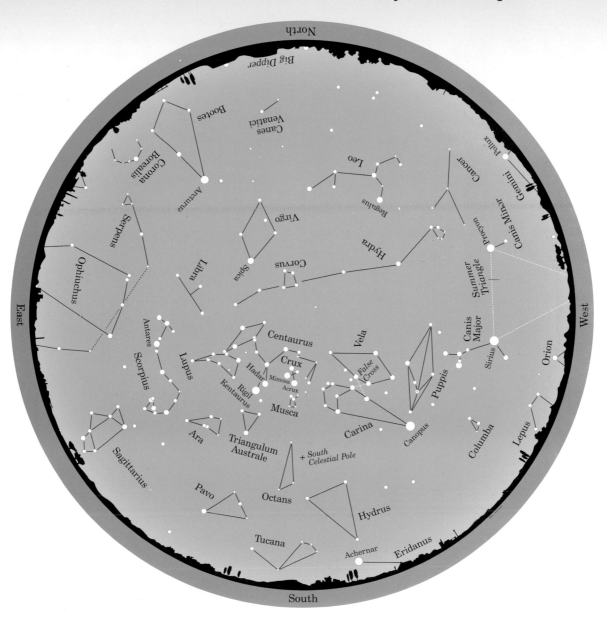

Winter Stars

June–July 10–12 pm
August–September 7–9 pm

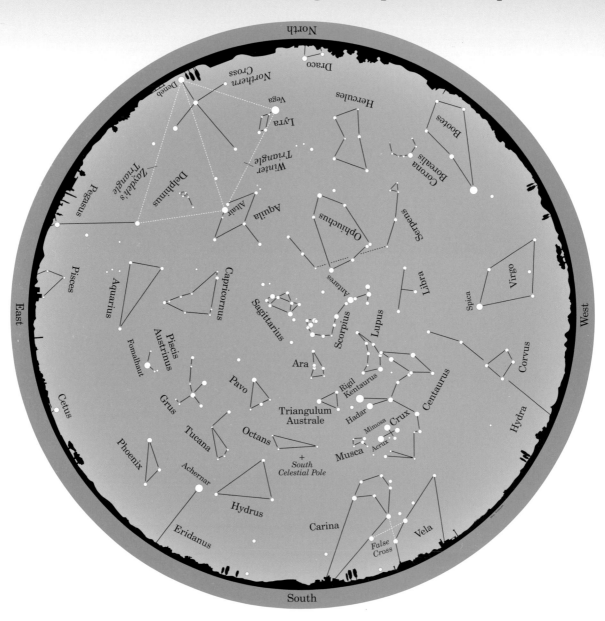

Part 2

A walk through the heavens

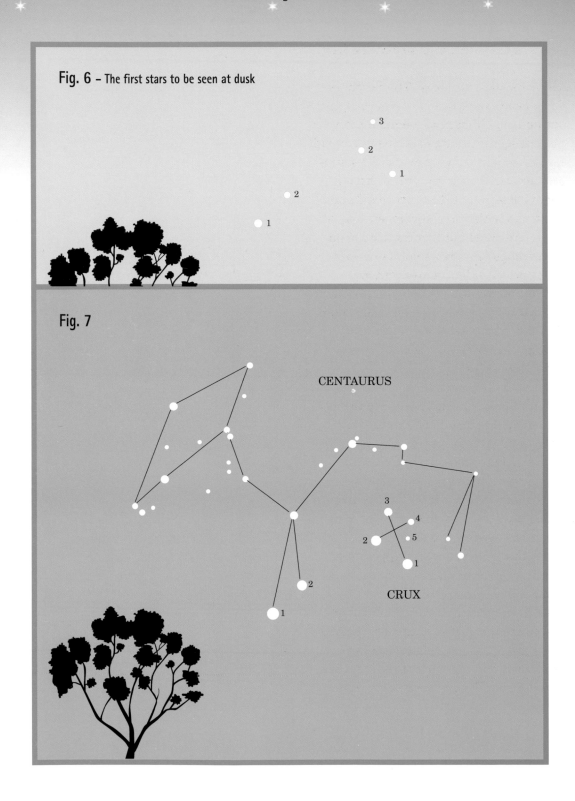

Fig. 6 – The first stars to be seen at dusk

Fig. 7

CENTAURUS

CRUX

A walk through the southern sky

The constellations Crux (The Southern Cross) and Centaurus are very conspicuous and may be seen throughout the year, but during the months of October through December they lie very low in the southern horizon. During that time Orion is very visible just north of overhead and remains visible from November to April. Therefore, we should concentrate upon both groups as starting points to walk from one constellation to another.

Since the written descriptions are from direct observation of the celestial sphere they are more accurate than the diagrams, which are flattened out versions of the celestial globe.

Let us start our walk by locating the constellations The Southern Cross (Crux), and Centaurus (the Centaur).

As the Sun goes down and the first stars appear you may note one or two bright stars that seem to be quite isolated in the sky. Ignore them at present. Face south and notice a five star group, of which Stars 1 and 2 which appear separate from the other three stars are especially bright. See Fig. 6.

As the night deepens and more stars appear this group becomes more defined. Stars 1 and 2 are the pointer stars of Centaurus and the other three stars are part of Crux (The Southern Cross). See Fig. 7.

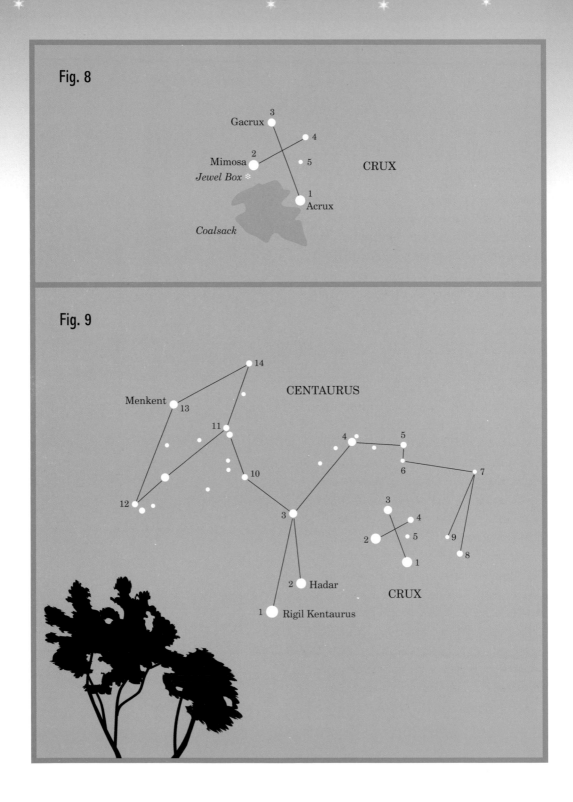

Fig. 8

Gacrux 3
Mimosa 2
Jewel Box
Coalsack
4
5
1 Acrux
CRUX

Fig. 9

CENTAURUS
14
Menkent 13
11
10
3
1 Rigil Kentaurus
2 Hadar
12
4 5
6 7
3
2 4
5
1
9
8
CRUX

The Southern Cross (Crux) (Fig. 8)

* Stars 1, 2 and 3 are brighter than Stars 4 and 5.

* Notice a hazy patch just below Star 2 of Crux. This is a globular cluster called the Jewel Box. It is best seen through binoculars.

* Just below the Jewel Box is an area of relative darkness due to the presence of a dark nebula in this region that hides the stars behind it. This area is called the Coalsack.

It is good to learn the names of some of the stars as we walk among them. They become much friendlier, so from now on I will refer to Star 1 of Crux by its proper name 'Acrux'.

When you are familiar with the Southern Cross you will note that a line from Star 4 through Star 2 leads to the pointer stars of Centaurus (Stars 1 and 2). See Fig. 9. Star 1, Rigil Kentaurus, or Alpha Kentaurus, is the closest star to Earth. It is 4.3 light years away.

At this time study the relationship between Centaurus and Crux until the visual image is clear. Notice that Centaurus, which vaguely resembles a camel with a diamond-shaped head, surrounds Crux on three sides.

Test you measuring ability. Star 8 of Centaurus is 20.5 degrees from Star 1 of Centaurus. Star 14 of Centaurus is 19 degrees from Star 12.

Read the legend of Centaurus.

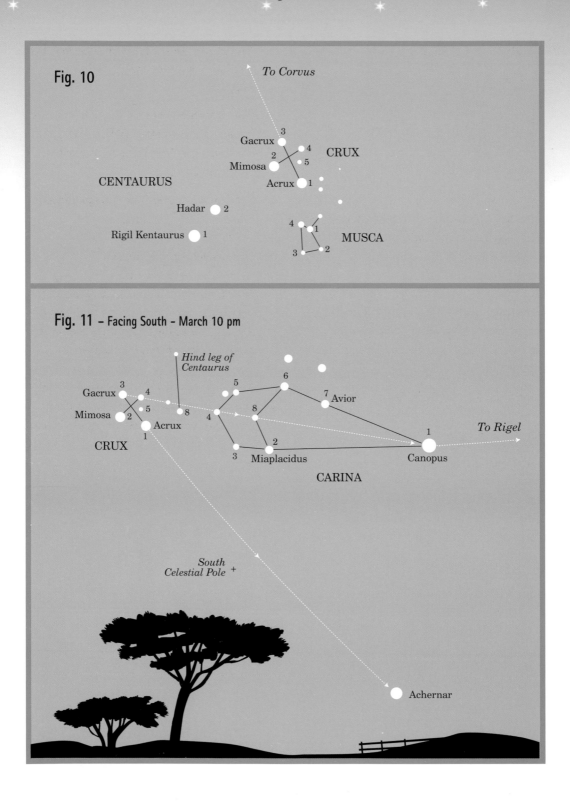

Fig. 10

To Corvus

CENTAURUS

Gacrux
Mimosa
Acrux

CRUX

Hadar 2

Rigil Kentaurus 1

MUSCA

Fig. 11 – Facing South – March 10 pm

Hind leg of
Centaurus

Gacrux
Mimosa
Acrux

CRUX

Avior

To Rigel

Miaplacidus

Canopus

CARINA

South
Celestial Pole +

Achernar

Pathways from Crux

To locate Musca (Fig. 10)

* Just below Crux is the dim constellation of Musca, the Fly.

To locate Corvus

* A line from Acrux (Star 1) through Star 3 of Crux leads straight to Corvus.

To locate Carina (Fig. 11)

* A line from Star 3 through Star 4 of Crux leads straight through the constellation of Carina and in a slight curve to Canopus, the second brightest star in the sky.

* A line from Star 2 of Crux through Acrux (Star 1) leads to Star 2 (Miaplacidus) of Carina and moves in a slight curve to Canopus.

* Carina looks like an ice cream cone with the top of the ice cream (Star 4) slightly off center.

* A line from Star 5 of Crux through Star 8 of Centaurus goes to Star 4 of Carina (tip of the ice cream), past Star 8 (top of the cone cracker) to Canopus.

* A line from Star 2 of Crux through Canopus leads straight to Rigel of Orion.

To locate Achernar of Eridanus

* A line from Star 3 of Crux through Acrux (Star 1) passes in a slight curve next to the South Celestial Pole to the bright star Achernar, the end of the heavenly river Eridanus.

At this time study the relationships between Canopus, Achernar and Crux and confirm your visualization of Carina.

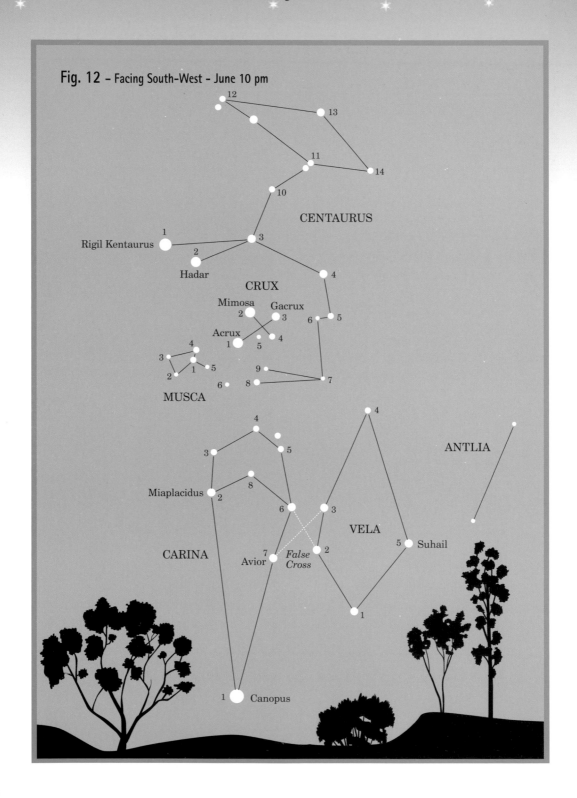

Fig. 12 – Facing South-West – June 10 pm

The Centaurus–Crux–Carina–Vela relationship
The True and False Southern Crosses

To locate Vela and the False Southern Cross (Fig. 12)

* Vela lies north of Carina between Centaurus and Canis Major. It is tied to Carina by a design that resembles Crux, the Southern Cross. It is therefore called the False Southern Cross. Some observers mistakenly think it is Crux.

* The False Cross is larger, but not as bright as Crux. It consists of Stars 3 and 2 of Vela and Stars 6 and 7 of Carina.

* Study the shape of Vela, which resembles a kite. Note that Stars 2 and 5 of Vela are wider apart than Stars 2 and 6 of Carina.

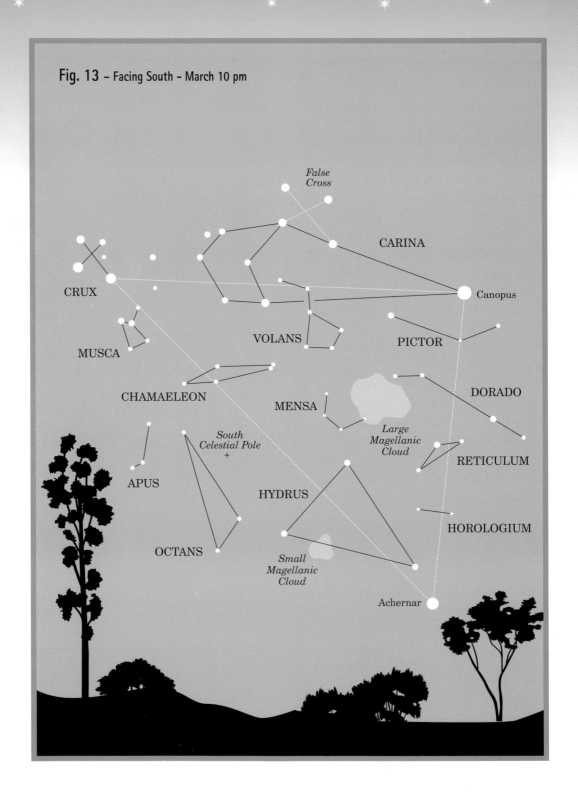

Fig. 13 – Facing South – March 10 pm

The Acrux–Canopus–Achernar triangle (Fig. 13). Best seen 9 pm December–April.

This triangle is remarkable in that it encloses two constellations and partially encloses eight others. The triangle contains the constellations of Mensa and Volans and partially encloses Musca, Chamaeleon, Carina, Pictor, Dorado, Horologium, Reticulum, and Hydrus. It also encloses the Large Magellanic Cloud.

Study this triangle to help visualize these constellations. They are very faint and unimportant except for Hydrus and Star 1 of Musca. Achernar is a bright star that is approximately four fully extended hand spans from Crux.

Notice the relationship between Octans and the South Celestial Pole.

The Brightnesses of the relevant constellations are:

Dorado – magnitude of 3.3 or dimmer
Chamaeleon – magnitude of 4.0 or dimmer
Mensa – magnitude of 5.1 or dimmer
Musca – magnitude of 2.7 or dimmer
Pictor – magnitude of 3.3 or dimmer
Reticulum – magnitude of 3.4 or dimmer
Volans – magnitude of 3.8 or dimmer
Apus – magnitude of 3.8 or dimmer
Octans – magnitude of 3.8 or dimmer
Hydrus – magnitude of 3.2 or dimmer
Horologium – magnitude of 5.0 or dimmer

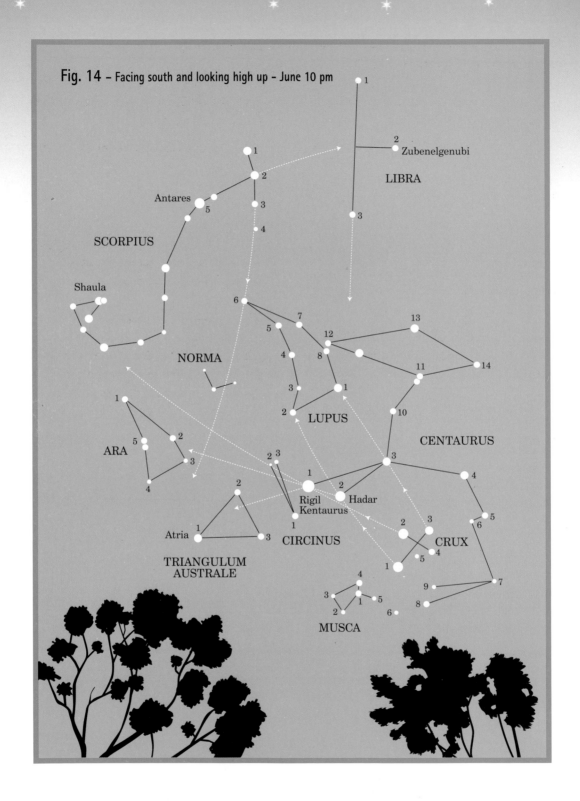

Fig. 14 – Facing south and looking high up – June 10 pm

The Centaurus–Lupus–Ara–Libra–Scorpius relationship (Fig. 14)

The name of Star 1 of Centaurus is Rigil Kentaurus. Star 2 is Hadar.

Pathways from Centaurus

To Lupus

* Lupus looks like a large rhinoceros horn. Star 12 of the head of Centaurus almost touches Star 8 of Lupus.

* Star 3 of Crux through Star 3 of Centaurus leads to Star 1 of Lupus.

* Star 1 of Crux through Star 2 of Centaurus is equidistant to and leads directly to Star 2 of Lupus.

* Star 2 of Crux through Star 1 (Rigil Kentaurus) of Centaurus leads to the tail of Scorpius. Follow the curve of the tail northward to Antares and to the four-star head of Scorpius.

* Follow the curve of the head of Scorpius to the tip of Lupus and continue past Norma in a curve to near the base of Ara, which resembles a small rhinoceros horn.

To locate Libra

* A line from Antares of Scorpius through Star 2 in the head of Scorpius leads directly to Libra, the Scale of Justice.

* The crossbar of Libra (the Scale) leads to the head of Centaurus.

To Ara

* Star 2 through Star 1 of Centaurus leads to Ara. It lies south of Scorpius and between Scorpius and Triangulum Australe.

To Circinus and Triangulum Australe

* A line from Star 3 of Centaurus through Rigil Kentaurus passes through the center of Circinus and continues to cross Triangulum Australe.

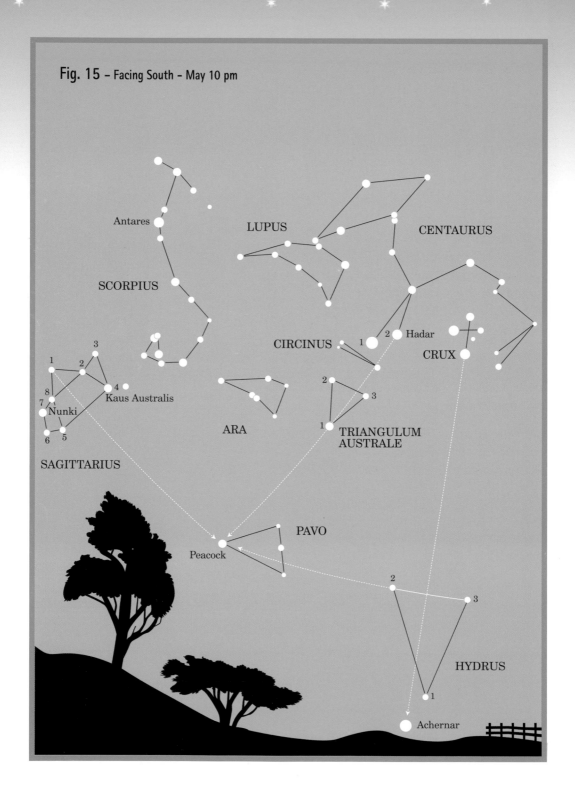

Fig. 15 – Facing South – May 10 pm

The Sagittarius–Centaurus–Hydrus– Pavo relationship (Fig. 15). Best seen 9 pm July (May–September)

The positions of the constellations are distorted in order to understand the pathways better.

To locate Peacock in Pavo

It is important to recognize the star Peacock in the constellation of Pavo. It serves as a good reference point, as a corner of a triangle that will later make it easy to locate the constellations of Grus, Tucana and Indus.

✳ A line from Hadar (Star 2) of Centaurus through Star 1 of Triangulum Australe goes directly to Peacock.

✳ A line from Star 3 of Triangulum Australe through Star 1 of the triangle leads to the Peacock area.

✳ A line from the top of the Sagittarius teapot (Star 1) moving through the base of the teapot (between Stars 4 and 5) leads directly to Peacock.

✳ A line from Star 3 of Hydrus through Star 2 of Hydrus leads to Peacock.

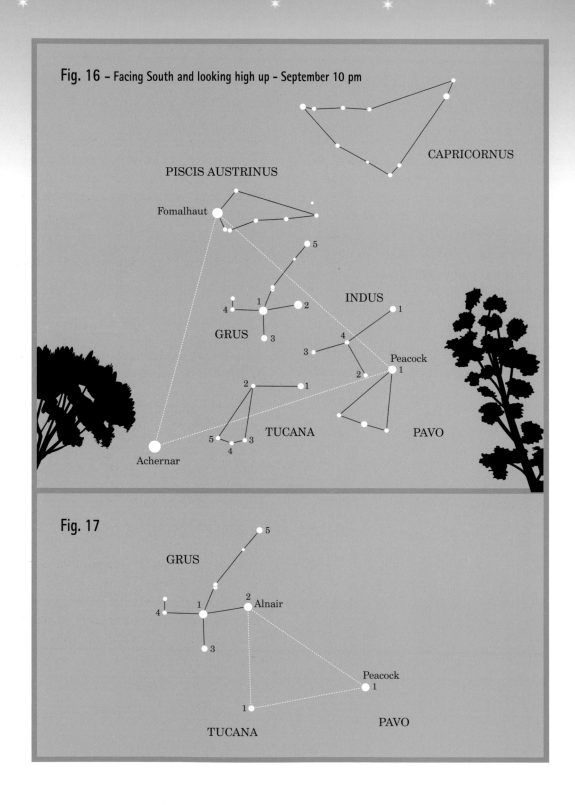

Fig. 16 – Facing South and looking high up – September 10 pm

CAPRICORNUS

PISCIS AUSTRINUS

Fomalhaut

INDUS

GRUS

TUCANA

Peacock

PAVO

Achernar

Fig. 17

GRUS

Alnair

Peacock

TUCANA

PAVO

The Achernar–Fomalhaut–Peacock triangle

To locate Grus (Fig. 16) Best seen 9 pm July–January

Now that you have identified Peacock of Pavo and Achernar of Eridanus, visualize a line between them, and look straight north to locate Fomalhaut of Piscis Austrinus. It is a bright star in a relatively dark area of the sky (see Fig. 34). Form a triangle joining them together. This triangle encloses Grus, and its lines cross Indus and Tucana.

One may draw a curve going from Fomalhaut through Grus through Pavo through Centaurus to the Southern Cross.

Pathways from Grus

To locate Tucana (Figs. 16 and 17)

∗ A line from Star 2 of Grus perpendicular to the bar 4–1–2 of Grus moving southward touches Star 1 of Tucana. If you then move at right angles from Tucana you reach the fairly bright star Peacock of Pavo. This almost forms a right-angle triangle.

∗ A line from Star 1 through Star 3 of Grus leads to Tucana.

To locate Indus (Fig. 16)

∗ A line from Star 4 of Grus through Stars 1 and 2 of Grus leads just above Star 1 of Indus.

∗ Star 2 of Indus is a faint star that is on a line between Star 1 of Tucana and Peacock.

∗ A line from Fomalhaut of Piscis Austrinus through Star 2 of Grus leads in a slight curve to Peacock.

To Locate Capricornus

∗ Follow a curve from Star 1 through Star 5 of Grus to Capricornus.

Before moving to the next section study the relationships formed by the triangle in Fig. 16.

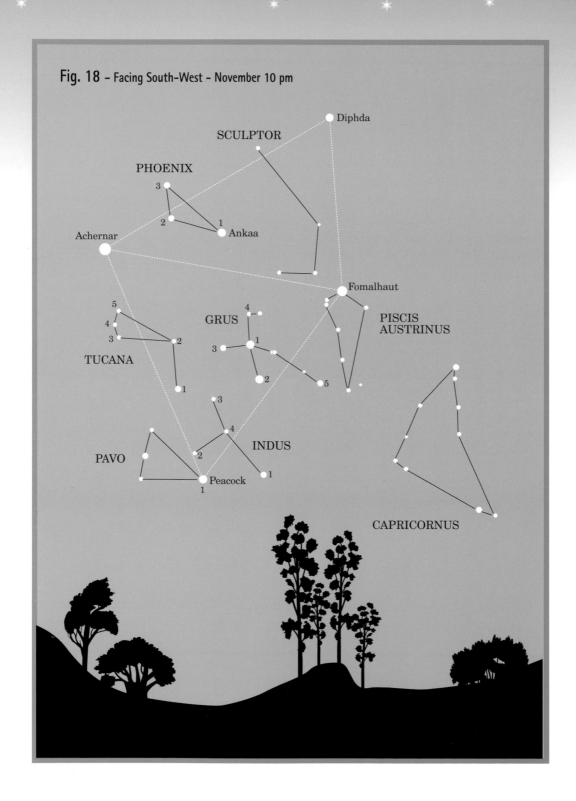

Fig. 18 – Facing South-West – November 10 pm

The Diphda–Fomalhaut–Achernar–Peacock triangles

To locate Ankaa in the constellation of Phoenix (Fig. 18)
Review the triangle formed by the stars Fomalhaut, Peacock and Achernar.

* Visualize a triangle joining Diphda of Cetus with Fomalhaut and Achernar. See Fig. 34 to see relationships of Diphda and Fomalhaut to Pegasus.

* This triangle encloses much of Phoenix, including its brightest star Ankaa, which lies in the approximate center of the triangle.

* Notice that the line from Achernar to Diphda crosses Star 2 of Phoenix.

* Star 2 through Star 4 of Grus leads to Ankaa, not quite equidistant.

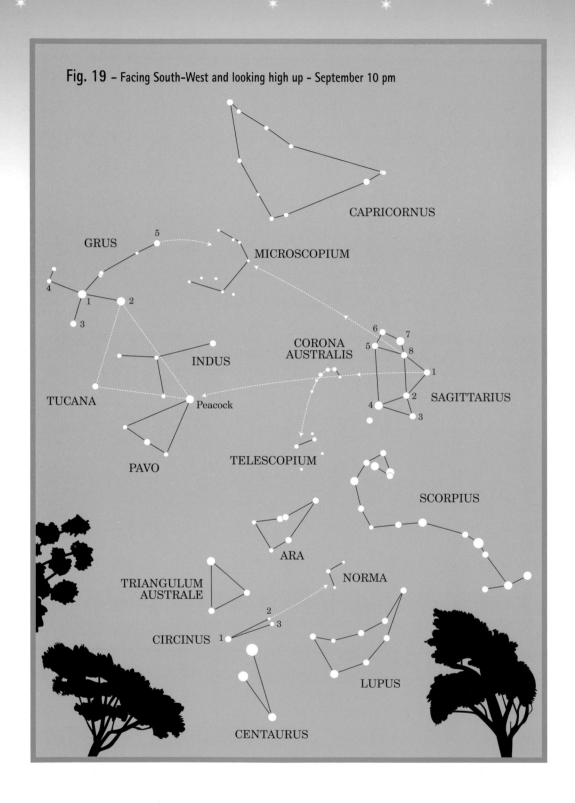

Fig. 19 – Facing South-West and looking high up – September 10 pm

CAPRICORNUS

GRUS

MICROSCOPIUM

TUCANA

INDUS

CORONA
AUSTRALIS

SAGITTARIUS

Peacock

PAVO

TELESCOPIUM

SCORPIUS

ARA

NORMA

TRIANGULUM
AUSTRALE

CIRCINUS

LUPUS

CENTAURUS

The Grus, Sagittarius, Scorpius and Lupus relationship (Fig. 19)

Study the above relationship as well as the four constellations Microscopium, Corona Australis, Norma and Telescopium. These constellation are very faint and unimportant.

To locate Microscopium

* Microscopium may be visualized as a wide bucket. Follow the curve of Grus from Star 1 through 5, and a sharp curve aims into the bucket.

* Star 1 through Star 8 of Sagittarius points to the bottom of the bucket.

To locate Telescopium

* The curve of Corona Australis points to Telescopium.

To locate Norma

* Norma is on a line with Star 1 and Star 2 of Circinus approximately 1.5 × the distance between Star 1 and Star 2.

Microscopium – magnitude of 4.7 or dimmer
Corona Australis – magnitude of 4.1 or dimmer
Norma – magnitude of 4.0 or dimmer
Telescopium – magnitude of 3.5 or dimmer

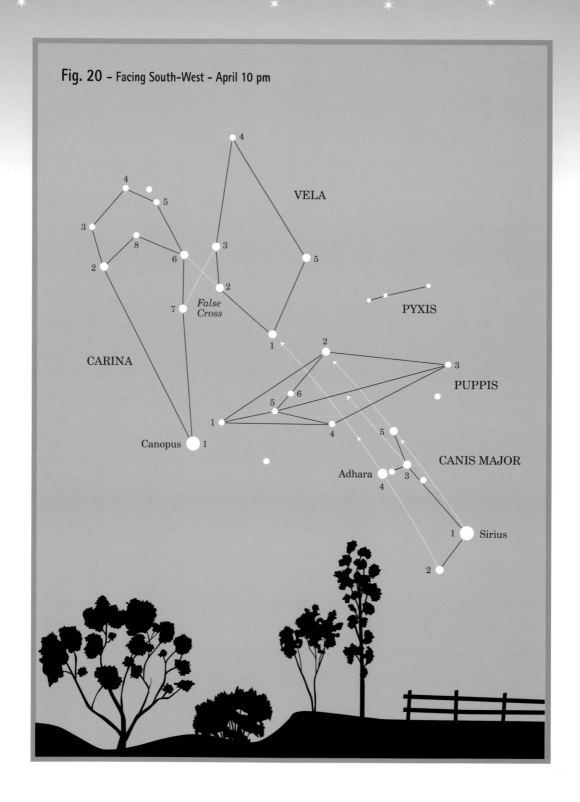

Fig. 20 – Facing South-West – April 10 pm

The Canis Major–Puppis–Carina (Canopus) relationship

To locate Puppis (Fig. 20)

∗ Puppis may be visualized as a long-faceted diamond that lies between Canis Major, Vela and Carina.

∗ A line from Sirius through Star 3 of Canis Major points to the center of Puppis.

∗ Star 1 of Canis Major, Sirius, through Star 5 leads to Star 2 of Puppis.

∗ A line from Sirius through Star 4 of Canis Major leads to Star 5 of Puppis.

∗ A line from Star 2 of Canis Major through Star 4 crosses Puppis to Star 1 of Vela.

∗ Notice that Star 1 of Puppis is close to Canopus, the second brightest star in the sky and that moving from Canopus through Star 1 of Puppis goes to Star 2 of Puppis.

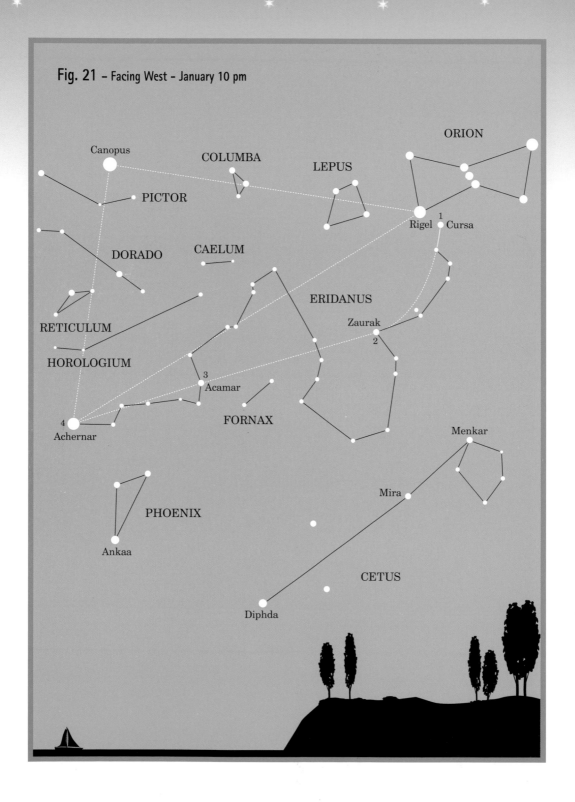

Fig. 21 – Facing West – January 10 pm

The River Eridanus (Fig. 21). Best seen 9 pm November to March

The complete extent of Eridanus lies directly overhead at approximately 9 pm January 1.

* Eridanus starts from the star Cursa, which lies next to Rigel of Orion. It then rambles southward to end at Achernar.

* The only bright stars in the constellation are Cursa (Star 1), Zaurak (Star 2), Acamar (Star 3) and Achernar (Star 4).

* It may be simpler to visualize the river by joining stars 1 2 3 4 to form a sinuous line from Cursa to Achernar.

Notice that Eridanus partially surrounds the faint constellation of Fornax.

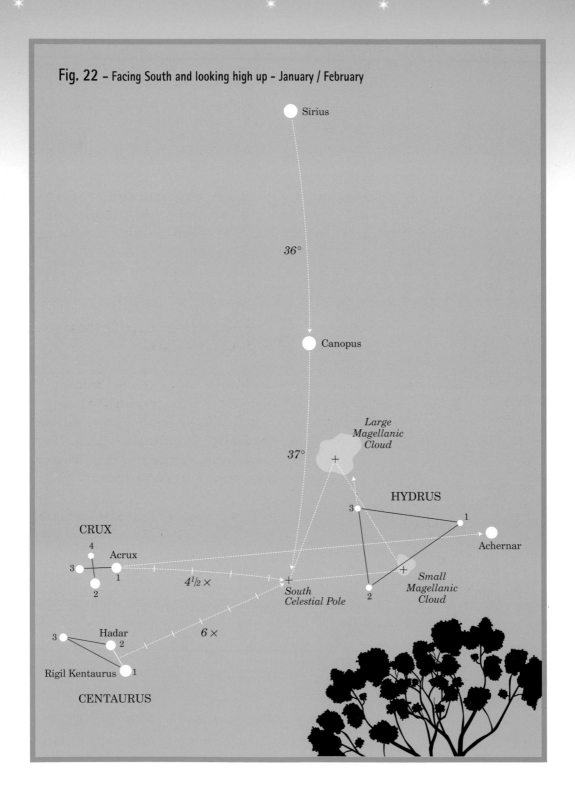

Fig. 22 – Facing South and looking high up - January / February

Sirius

36°

Canopus

Large Magellanic Cloud

37°

HYDRUS

3

1

Achernar

CRUX

4

Acrux

3

1

2

4½ ×

+
South Celestial Pole

+
2

+
Small Magellanic Cloud

3

Hadar

2

6 ×

Rigil Kentaurus 1

CENTAURUS

Guides to the South Celestial Pole (SCP) (Fig. 22, not to scale)

* The closest constellation to the South Celestial Pole is the dim constellation Octans. It is too dim to be of much value in locating the SCP.

* The South Celestial Pole lies midway between Acrux (Star 1 of Crux) and Achernar of Eridanus, but 4.5 degrees to the side of that line.

* Moving along the line from Star 3 of Crux through Acrux, approximately 4½ times the length of Crux leads to the South Celestial Pole.

* The South Celestial Pole is on a line perpendicular to the midpoint of a line between Rigil Kentaurus (Star 1) and Hadar (Star 2) of Centaurus. The distance to the SCP is approximately six times the distance between Rigil Kentaurus and Hadar.

* If you flip the triangle of Hydrus over, Star 1 of Hydrus, which is closest to Achernar, will lay near the South Celestial Pole.

* Star 2 of Hydrus is 13 degrees from the South Celestial Pole.

* A line from Sirius to Canopus leads to the South Celestial Pole. The distance between Sirius and Canopus is 36 degrees, between Canopus and the SCP is 37 degrees – almost equidistant.

* Canopus and Achernar form an almost equilateral triangle with the SCP.

* The Small Magellanic Cloud (SMC) and the Large Magellanic Cloud (LMC) form a nice equilateral triangle with the South Celestial Pole.

The Magellanic Clouds (best seen September–March)

The Magellanic Clouds, or Clouds of Magellan, originally called the Cape Clouds, were discovered before the time of the explorer Ferdinand Magellan. They are companion galaxies to our own galaxy and appear as fuzzy patches.

* Star 2 through Star 3 of Hydrus leads to the LMC. The line joining Stars 1 and 2 of Hydrus crosses the SMC.

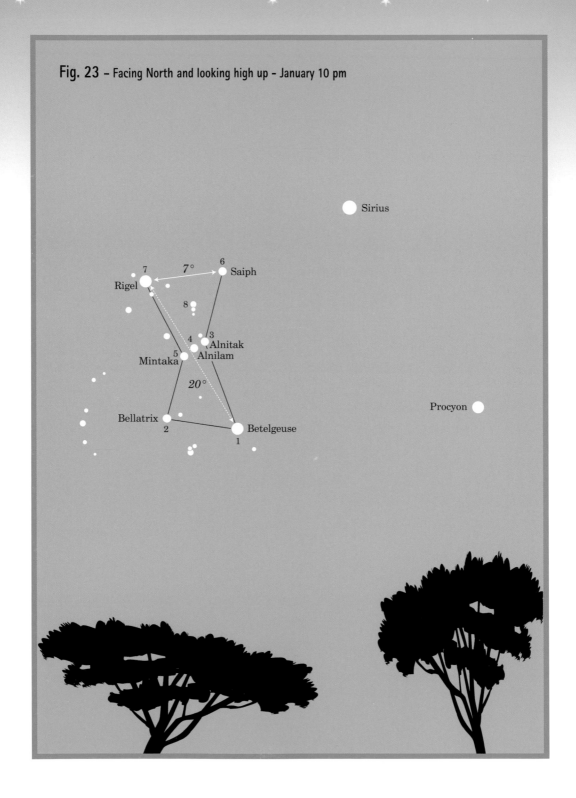

Fig. 23 – Facing North and looking high up – January 10 pm

Starting from Orion (Fig. 23)

Orion, the Hunter, is a large striking constellation that may be seen in the northern sky during the summer months December to April. Look for seven bright stars in the shape of an hourglass with three bright stars (Stars 3, 4, 5) in a row forming the waist of the hourglass or the belt of the Hunter. Star 1 is Betelgeuse and Star 7 is Rigel.

Once you have identified Orion, test your ability to measure distances. Stars 6 and 7 are approximately 7 degrees apart. The distance between Stars 1 and 7 is 20 degrees.

Study Orion closely since it serves as an excellent guide to the surrounding constellations and to the Pleiades.

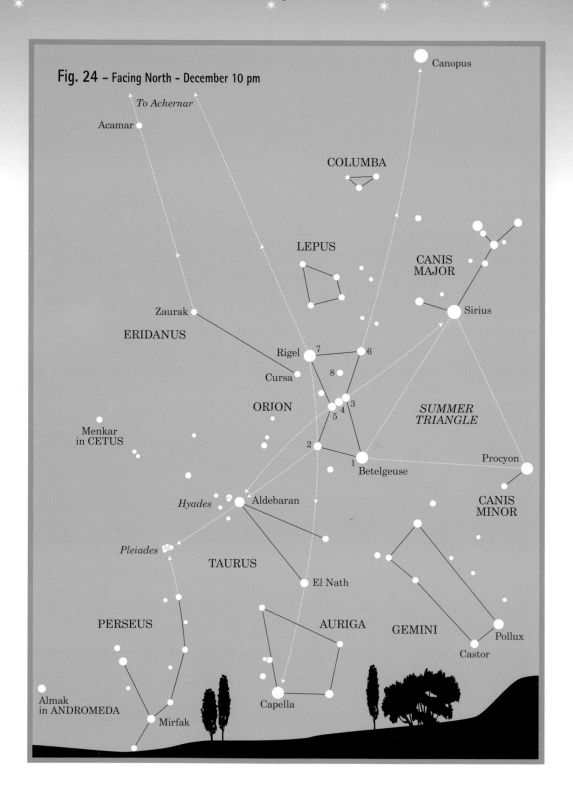

Fig. 24 – Facing North – December 10 pm

Pathways from Orion

To locate Sirius in Canis Major (Fig. 24)
Sirius is the brightest star in the sky.

* A line through the belt of Orion (Stars 3, 4, 5), curves slightly southward to Sirius on one side and slightly northward to Aldebaran of Taurus on the other side. Sirius is about 15 degrees from Star 6 (Saiph) of Orion. Aldebaran is 15 degrees from Star 2 (Bellatrix) of Orion.

To locate Procyon in Canis Minor
This is a bright star 25 degrees in a northerly direction from Sirius. You may notice that the Milky Way lies between Procyon and Sirius.

The Summer Triangle (Winter Triangle in the Northern Hemisphere) is an equilateral triangle joining Betelgeuse of Orion, Sirius of Canis Major and Procyon of Canis Minor.

To locate Taurus and its star clusters the Hyades and the Pleiades

* A line from Rigel (Star 7) through Bellatrix (Star 2) of Orion leads directly to El Nath of Taurus and then almost in a straight line to Capella in Auriga.

* From Orion walk past Aldebaran approximately 10 degrees to a faint hazy cluster of stars, the Pleiades.

* Move from Mirfak of Perseus and then southward along the curve of Perseus to the Pleiades.

To locate Achernar and Canopus

* A line from Star 5 of Orion moving through Rigel (Star 7) of Orion leads in a straight line to Achernar.

* A line from Star 3 through Star 6 of Orion passes between Lepus and Canis Major to Canopus.

Read about the Pleiades in the legends of Taurus and Ursa Major.

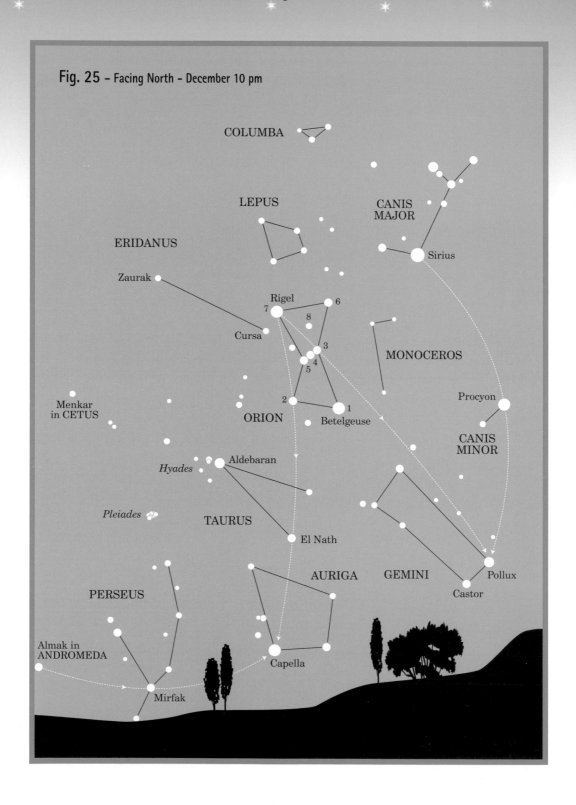

Fig. 25 – Facing North – December 10 pm

To locate Gemini (Fig. 25)
Gemini forms a nice rectangle that
includes the stars Pollux and Castor,
which are frequently referred to as the
Twins. They symbolize true friendship.

* A curve from Sirius to Procyon leads to Pollux
  and Castor.

* A line from Star 7 (Rigel) of Orion through
  Star 3 leads to Pollux.

To locate Auriga
Auriga, the Charioteer, is shaped like a
kite. It lies north of the horns of Taurus.
Capella is the sixth brightest star in the
sky. Just below Capella are three faint
stars, called the Kids (baby goats).

* An arc from Almak of Andromeda through
  Mirfak of Perseus leads to the bright star
  Capella.

* A walk from Star 4 of Orion going between
  Star 2 and Star 1, passes through El Nath of
  Taurus to Capella.

  Capella is 23 degrees from Aldebaran
and from Castor. It is 10 degrees from
Mirfak and 35 degrees from Betelgeuse.
Test your ability to measure these distances.

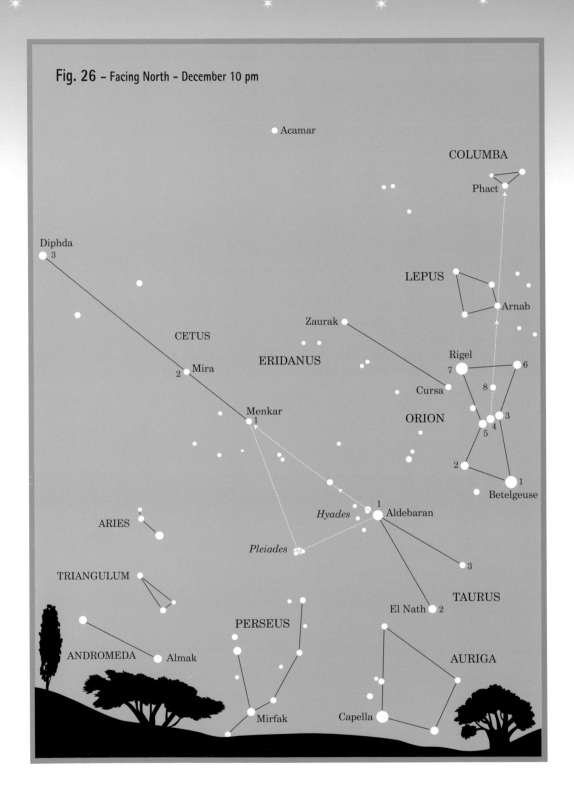

Fig. 26 – Facing North – December 10 pm

To locate Lepus, Columba and Cetus (Fig. 26)

* A line from Star 4 of Orion through Star 8 leads to Arnab of Lepus. This line continues to Phact of Columba.

* Stars 6 and 7 of Orion form a triangle with Arnab in Lepus.

To locate Cetus

This constellation has three fairly bright stars, Menkar, Star 1, Mira, Star 2 (which is a variable star; its brightness fluctuates), and Diphda, Star 3.

* A large right triangle is formed by joining the Pleiades, Aldebaran of Taurus and Menkar of Cetus.

* The tip of the southern horn of Taurus (Star 3) through Aldebaran (Star 1) of Taurus curves slightly to Menkar in Cetus – 25 degrees from Aldebaran.

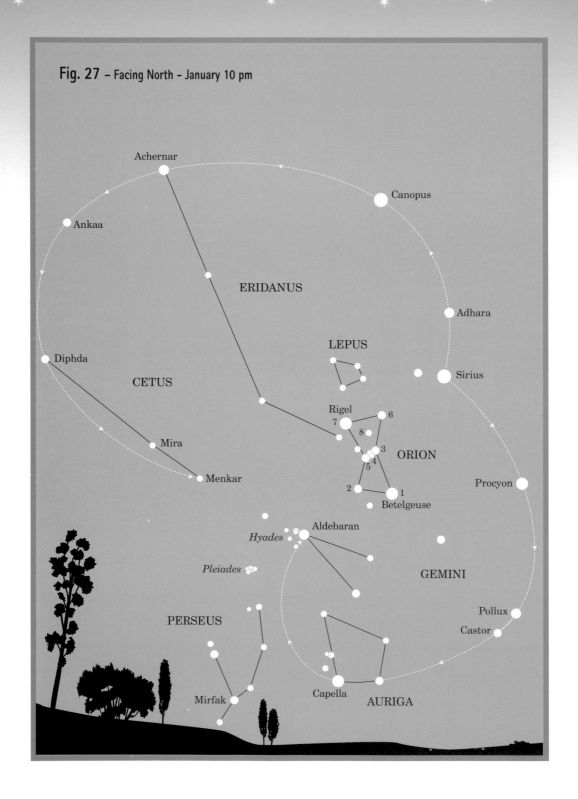

Fig. 27 – Facing North – January 10 pm

Orion's encirclement (Fig. 27)

* *The northern area* of Orion is surrounded by
  five very bright stars. Starting from Betelgeuse
  (Star 1) of Orion move to
      Procyon of Canis Minor and then in a curve
      through
      Pollux of Gemini,
      Castor of Gemini,
      Capella of Auriga, and then to
      Aldebaran of Taurus.

* *The southern area* of Orion is surrounded by
  six prominent stars. Starting from the belt of
  Orion move in a curve to
      Sirius of Canis Major,
      Canopus of Carina,
      Achernar of Eridanus,
      Ankaa of Phoenix,
      Diphda of Cetus, and
      Menkar of Cetus.

* Solidify a mental image of the stars encircling
  Orion.

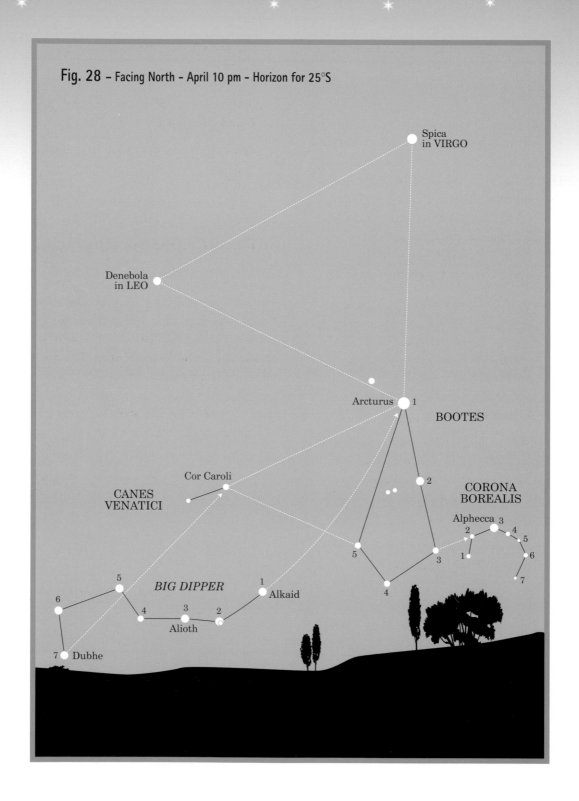

Fig. 28 – Facing North – April 10 pm – Horizon for 25°S

Spica
in VIRGO

Denebola
in LEO

Arcturus 1

BOOTES

2

Cor Caroli

CANES
VENATICI

CORONA
BOREALIS

Alphecca 3
2   4
5
1   6
5   3   7

5   4

BIG DIPPER   1 Alkaid

6   4   3   2

Alioth

7 Dubhe

To locate Arcturus in Bootes (Fig. 28). Best seen 9 pm May 1 to August 1

Bootes looks like a giant kite. In 1933 the light from Arcturus, the fourth brightest star in the sky, which is approximately 36 light years away, was focused onto a photo-electric cell which produced electrical current. This current was amplified and then used to open the gate of the 1933 World's Fair in Chicago Illinois, USA. That light had left Arcturus in 1897!

* Follow the curve of the handle of Ursa Major to Arcturus, the first bright star along that curve.

* A line from Antares of Scorpius through Star 1 of Scorpius leads in a very slight curve to Arcturus. See Fig. 29.

* Using Fig. 29, follow a line from Acrux (Star 1) of Crux through Star 2 of Crux leads across the sky to the very bright star Arcturus near the northern horizon.

* Form an equilateral triangle joining Denebola in Leo, Spica in Virgo and Arcturus.

To locate Cor Caroli in Canes Venatici

* A line from Star 7 of Ursa Major going between Stars 4 and 5 leads to Cor Caroli.

* Stars 1 and 5 of Bootes form a triangle with Cor Caroli.

To locate Corona Borealis

Corona Borealis, the Northern Crown, is a charming constellation. It lies between Bootes and Hercules, but is closer to Bootes (see Fig. 37). Alphecca, Star 3, is called The Jewel in the Crown. The six other stars are very faint.

* Move from Arcturus to Star 3 of Bootes. Alphecca is just to the side of this star.

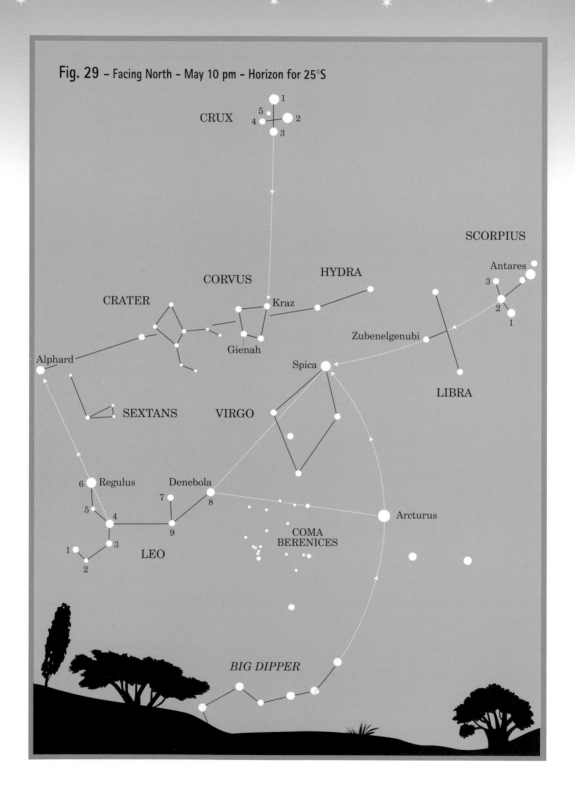

Fig. 29 – Facing North – May 10 pm – Horizon for 25°S

To locate Spica in Virgo (Fig. 29)

* Follow the arc of the handle of Ursa Major past Arcturus and continue the arc to a bright star which is Spica. Find the irregular diamond shape of Virgo.

* A line from Antares of Scorpius through Star 2 of Scorpius goes directly through Libra in a curve to Spica.

* Spica is part of an equilateral triangle with Arcturus, and Denebola of Leo.

To locate Corvus

* Follow the curve of the handle of Ursa Major to Arcturus continuing past Spica to locate Kraz, which is Star 1 in Corvus.

* A line from Star 1 of Crux through Star 3 leads to Corvus and continues to Denebola of Leo.

To locate Hydra, the Water Snake

This serpent, which was slain by Hercules, is the longest constellation in the sky, 100 degrees long. During autumn it extends across the sky south of Virgo, Corvus, Leo and Cancer. It actually cuts across Crater and Corvus. There is only one bright star – Alphard.

* A line from Star 4 of Leo through Star 6 (Regulus) leads directly to Alphard.

53

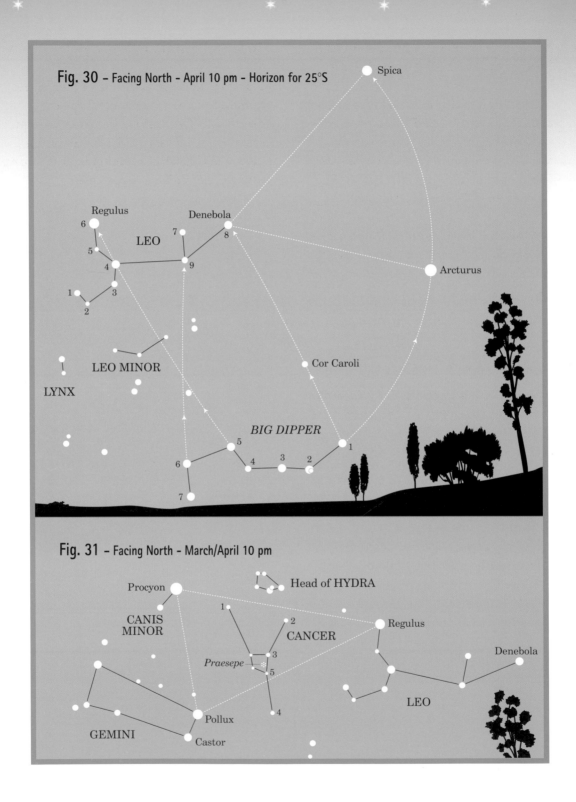

Fig. 30 – Facing North – April 10 pm – Horizon for 25°S

Spica

Regulus
6
Denebola
5
7
8
LEO
4
9
Arcturus
1
3
2

LEO MINOR

LYNX

Cor Caroli

BIG DIPPER
5
6
4
3
2
1
7

Fig. 31 – Facing North – March/April 10 pm

Procyon
Head of HYDRA

CANIS
MINOR
1
2
CANCER
Regulus
Denebola
Praesepe
3
5
LEO
4
Pollux
GEMINI
Castor

To locate Leo (Fig. 30). Best seen 9pm March 1 to 9 pm mid-June

Leo looks like a toy horse. The head, chest and front legs look like a backwards question mark. There are two stars in Leo whose names you should learn. Star 6 is Regulus, which means Little King. The tip of Leo's tail, Star 8, is Denebola.

* A line from Acrux, Star 1 of Crux through Star 3 of Crux leads through Corvus and then to Denebola, the tail of Leo. See Fig. 29.

* Form a triangle joining Arcturus of Bootes, Spica of Virgo and Denebola of Leo.

* A line from Betelgeuse of Orion to Procyon of Canis Minor leads in a slight curve to Regulus.

* Leo lies under the bowl of Ursa Major and lies inside the curve formed by the handle of the Ursa Major, Arcturus, Spica and Kraz in Corvus. See Fig. 29.

* A line from Star 4 of Ursa Major through Star 5 crosses the neck of Leo to Regulus. See Fig. 30.

* Form a triangle joining Pollux of Gemini, Procyon of Canis Minor and Regulus. See Fig. 31.

To locate Cancer (Fig. 31)

The very faint constellation of Cancer lies in the center of the triangle formed by Procyon of Canis Minor, Pollux of Gemini and Regulus of Leo. In the center of the square of Cancer is a faint cluster of stars called the Beehive or Praesepe.

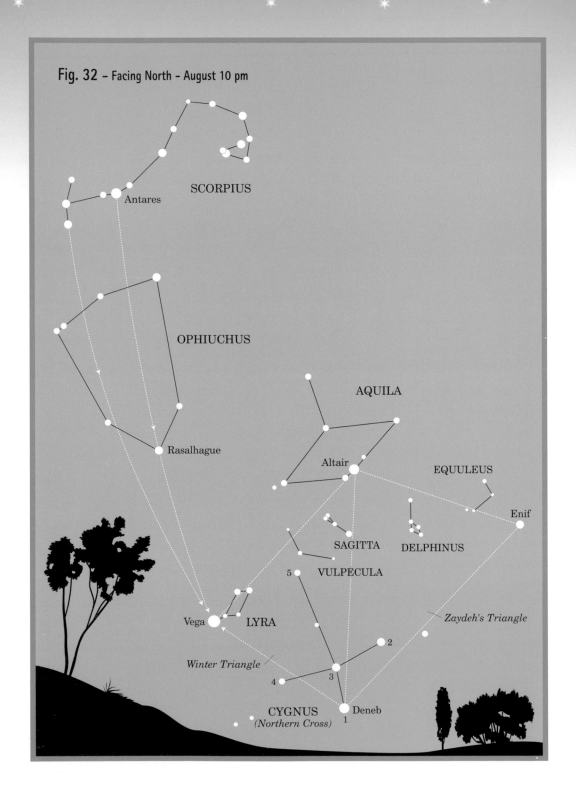

Fig. 32 – Facing North – August 10 pm

SCORPIUS

Antares

OPHIUCHUS

Rasalhague

AQUILA

Altair

EQUULEUS

Enif

SAGITTA

DELPHINUS

VULPECULA

5

Vega LYRA

Zaydeh's Triangle

Winter Triangle

4   3   2

CYGNUS
(Northern Cross)   Deneb
1

To locate Vega in Lyra (Fig. 32)

Vega is one of the brightest stars in the northern sky (magnitude 0.3).

* Go from Deneb (Star 1 of Cygnus) in a slight curve past Star 4 to the very bright star Vega.

* A line from Antares of Scorpius through Rasalhague of Ophiuchus leads across the sky in a mild curve to Vega.

* Follow the curve of the head of Scorpius through Ophiuchus to Vega near the northern horizon.

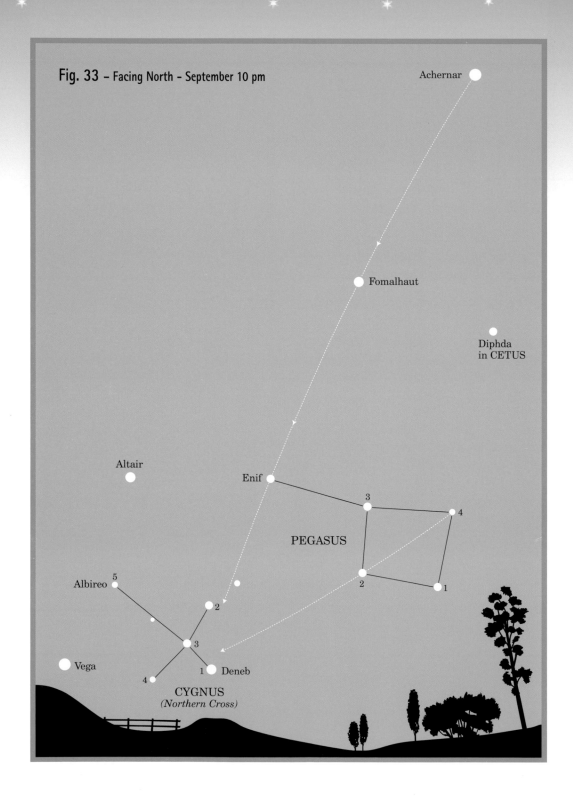

Fig. 33 – Facing North – September 10 pm

To locate the Northern Cross (Cygnus) (Fig. 33). Best seen 9 pm mid August to October

Cygnus, the Swan, lies along the Milky Way near the northern horizon. Star 1 (Deneb), a supergiant, is 60 000 times brighter than the Sun, but is 1500 light years away.

* A line from Star 4 of Pegasus through Star 2 leads to the area of Star 1 and Star 3 of Cygnus.

* A mild curve from Achernar through Fomalhaut through Enif of Pegasus leads across the sky to Cygnus at the northern horizon.

Study the Northern Cross and Vega relationship.

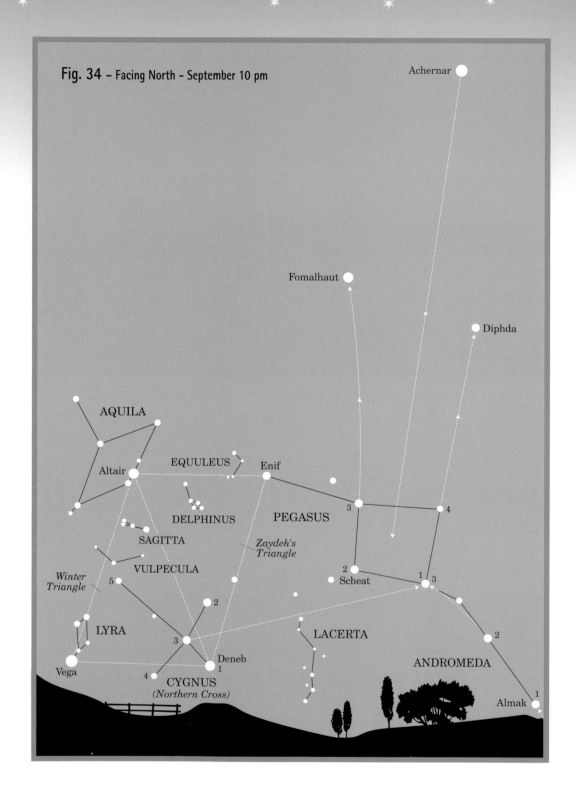

Fig. 34 – Facing North – September 10 pm

To locate Pegasus (Figs. 34 and 35)

There are several ways to find the giant square of Pegasus, the winged horse.

* A line from Achernar going between Diphda of Cetus and Fomalhaut of Piscis Austrinus leads to Pegasus near the northern horizon.

* We should be aware of the **Winter Triangle** This important star relationship is formed by Deneb of Cygnus (the Northern Cross), Vega of Lyra, and a third bright star, Altair of Aquila.

* Join Deneb, Altair and Enif to form Zaydeh's triangle opposite the Winter Triangle. This tri-angle is an easy way to identify Enif and then to move from Altair through Enif in an mild curve to the square of Pegasus.

* A line through the crossbar of Cygnus points in a slight curve to Enif in Pegasus.

* A line from Star 3 of Cygnus between Deneb and Star 2 goes directly to Star 1 of Pegasus.

* Follow a line in a soft curve from Mirfak in Perseus (see Fig. 26) to Almak of Andromeda past Star 2 to Star 3 of Andromeda, which is the same as Star 1 of Pegasus. This is one corner of the square of Pegasus.

To locate Altair in Aquila

* A line from Deneb going between Stars 2 and 3 of Cygnus leads to Altair in Aquila. Aquila is a diamond-shaped constellation with a tail at the corner of the diamond opposite Altair. Altair is its only bright star. Look closely at Altair; notice the faint stars on either side of it.

Study the **Winter Triangle** (see Fig. 34), which contains the constellations of Sagitta and Vulpecula. Zaydeh's triangle contains Delphinus and meets Equuleus.

Study these relationships to get a good image of the giant square of Pegasus, which is over 15 degrees wide.

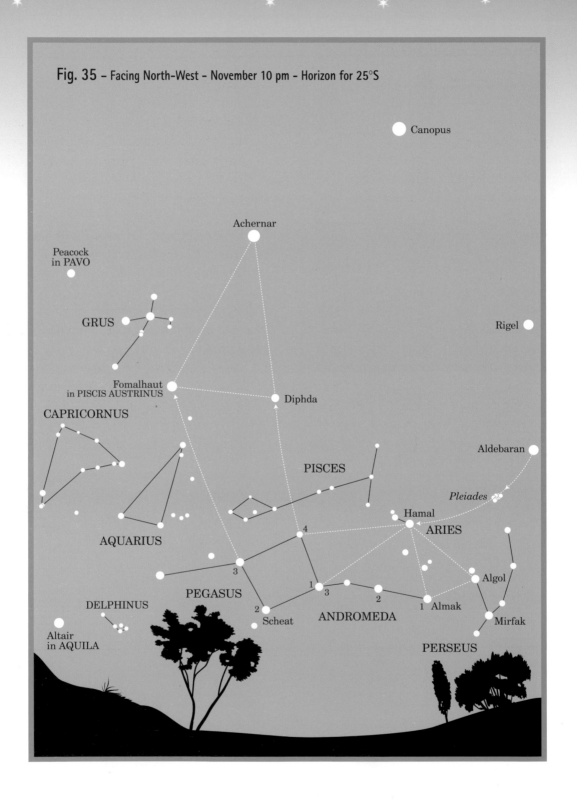

Fig. 35 – Facing North-West – November 10 pm – Horizon for 25°S

To locate Hamal in Aries, Diphda in Cetus and Fomalhaut in Piscis Austrinus (Fig. 35)

To locate Aries
From Perseus:

* Algol of Perseus and Almak of Andromeda form a triangle with the star Hamal of Aries.

From the square of Pegasus:

* Star 1 and Star 4 of Pegasus, when joined to Hamal of Aries, form large isosceles triangle.

From Orion:

* Find Aldebaran in Taurus (see Fig. 26) then move through the Pleiades and continue in a soft curve southward approximately 25 degrees to the moderately bright star Hamal of Aries.

To locate Fomalhaut in Piscis Austrinus
Fomalhaut is one of the brightest isolated stars in the southern sky.

* A line from Star 2 (Scheat) of Pegasus through Star 3 (Markab) leads to Fomalhaut, which is approximately 50 degrees from Star 3.

To locate Diphda in Cetus

* A line from Star 1 of Pegasus through Star 4 leads to Diphda, which is to the side of Fomalhaut.

An important orienting triangle is formed by joining Diphda, Fomalhaut and Achernar.

Study the relationship between Pegasus, Pisces, Aquarius, Capricornus, Delphinus and Aquila.

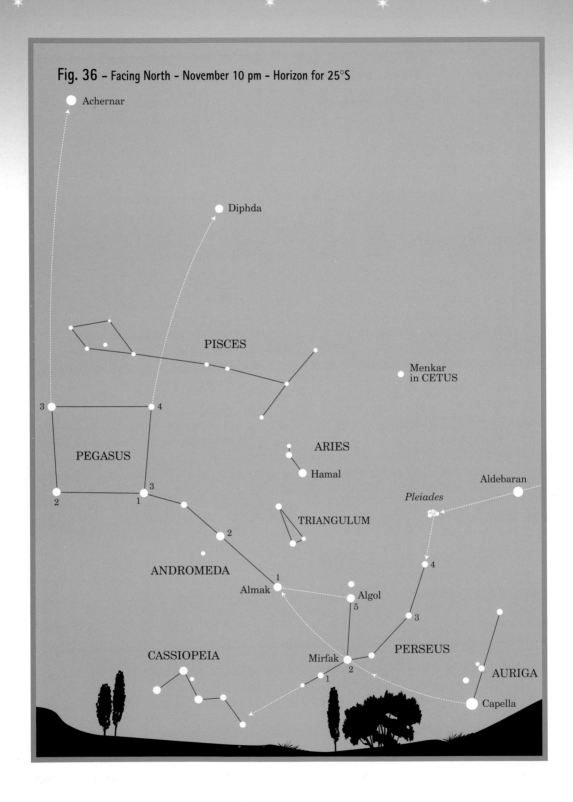

Fig. 36 – Facing North – November 10 pm – Horizon for 25°S

To locate Perseus and Andromeda (Fig. 36)

Perseus crosses the Milky Way. It may be seen on the extreme northern horizon.

* Using Fig. 27, follow the northern circle of stars over Orion from Procyon to Pollux to Castor to Capella and then instead of curving southward to Aldebaran, continue in a soft curve to Mirfak in Perseus and to Almak of Andromeda. See Fig. 36.

* Algol (Star 5 of Perseus) almost forms a right-angle triangle with Mirfak and Almak of Andromeda. Mirfak and Algol are the two brightest stars in Perseus.

* Using Fig. 25, from the belt of Orion move to Aldebaran and to the Pleiades. Just northerly of the Pleiades is the curve of stars of Perseus leading to Mirfak.

Now visualize the curve of five or six stars of Perseus which leads to the Pleiades going south, or to Cassiopeia going north. Cassiopeia may be below the northern horizon.

Andromeda lies between Perseus and Pegasus. It contains the only major galaxy outside of our own galaxy that we can see with our naked eye.

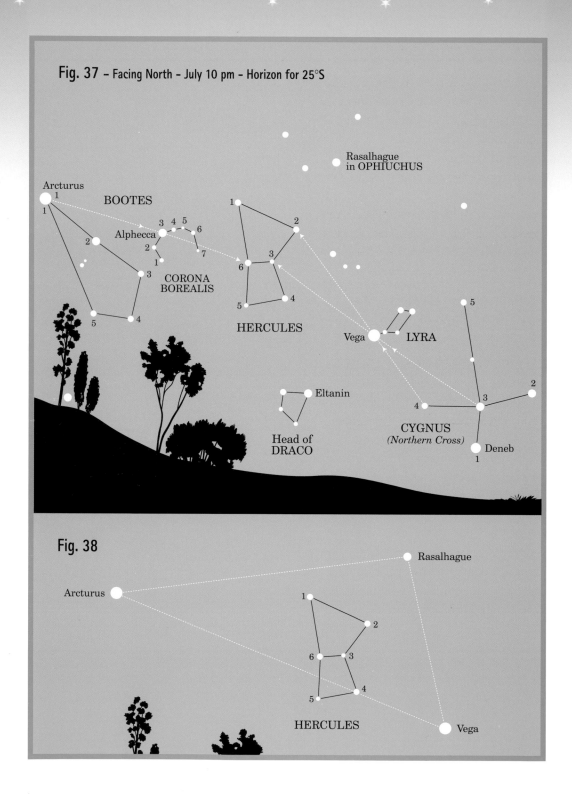

Fig. 37 – Facing North – July 10 pm – Horizon for 25°S

Rasalhague
in OPHIUCHUS

Arcturus

BOOTES

Alphecca

CORONA
BOREALIS

HERCULES

Vega   LYRA

Eltanin

Head of
DRACO

CYGNUS
(Northern Cross)

Deneb

Fig. 38

Rasalhague

Arcturus

HERCULES

Vega

To locate Hercules (Fig. 37)

Hercules is a faint constellation near the northern horizon that lies between Lyra and Corona Borealis. Most of Hercules lies within a giant triangle formed by Vega of Lyra, Arcturus of Bootes and Rasalhague of Ophiuchus. See Fig. 38.

* A line from Star 4 of Cygnus through Vega goes to Star 2 of Hercules. This distance is about 1½ times the distance between the tip of the crossbar of Cygnus and Vega.

* A line from Star 3 of Cygnus through Vega goes to Star 3 of Hercules.

* From Arcturus through Alphecca, the bright star of Corona Borealis (Star 3), to Star 6 of Hercules is a straight line.

Now construct the bent hourglass shape of the six stars of Hercules.

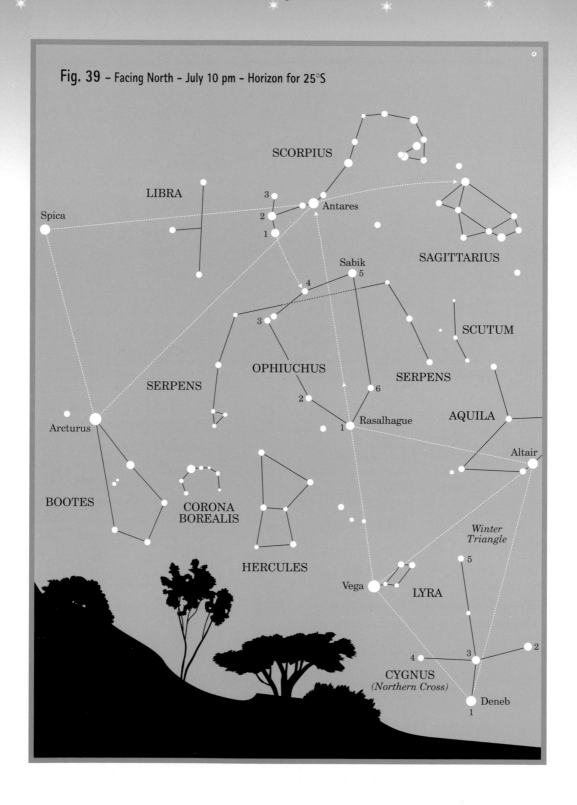

Fig. 39 – Facing North – July 10 pm – Horizon for 25°S

To locate Ophiuchus (Fig. 39)

Ophiuchus is known as the Serpent Bearer.

* On the south side of the Winter Triangle form an equilateral triangle by joining Altair of Aquila, Vega of Lyra and Rasalhague (Star 1) of Ophiuchus.

* The arc of Scorpius 3–2–1 leads to Star 4 of Ophiuchus.

Rasalhague (Star 1) is about 30 degrees from Sabik (Star 5).

To locate Serpens

Serpens is a divided constellation which appears to pass through the lower part of Ophiuchus. Its head and tail point northward.

To locate Antares in Scorpius

Scorpius is a bright constellation that actually resembles a large scorpion. Its brightest star, Antares, is flanked by a star on either side.

* From the north a line from Vega of Lyra crosses the sky to Rasalhague of Ophiuchus and then goes in a mild curve to Antares.

* A large triangle is formed by joining Arcturus of Bootes, Spica of Virgo and Antares.

* A line from Acrux (Star 1 of Crux) through Hadar (Star 2 of Centaurus) leads to Antares. See Fig. 14.

* The nose of Centaurus points to Scorpius. See Fig. 14.

* Review Fig. 14.

To locate Sagittarius

Sagittarius, the Teapot, lies between our solar system and the center of our galaxy. The Milky Way is widest and most dense behind Sagittarius.

* A line from the mid star of the head of Scorpius (Star 2) through Antares leads to the bottom star of Sagittarius. See also Fig. 15.

Study relationships between Altair of Aquila, Scorpius, Sagittarius, Cygnus, Vega of Lyra, Arcturus of Bootes and Spica of Virgo.

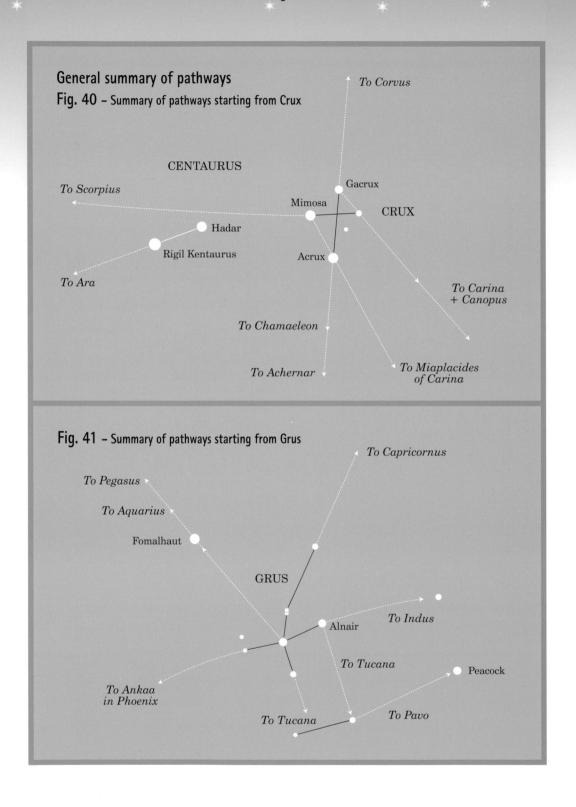

General summary of pathways
Fig. 40 – Summary of pathways starting from Crux

CENTAURUS

To Corvus

To Scorpius

Gacrux

Mimosa

CRUX

Hadar

Rigil Kentaurus

Acrux

To Ara

To Chamaeleon

To Carina
+ Canopus

To Achernar

To Miaplacides
of Carina

Fig. 41 – Summary of pathways starting from Grus

To Capricornus

To Pegasus

To Aquarius

Fomalhaut

GRUS

Alnair

To Indus

To Tucana

Peacock

To Ankaa
in Phoenix

To Tucana

To Pavo

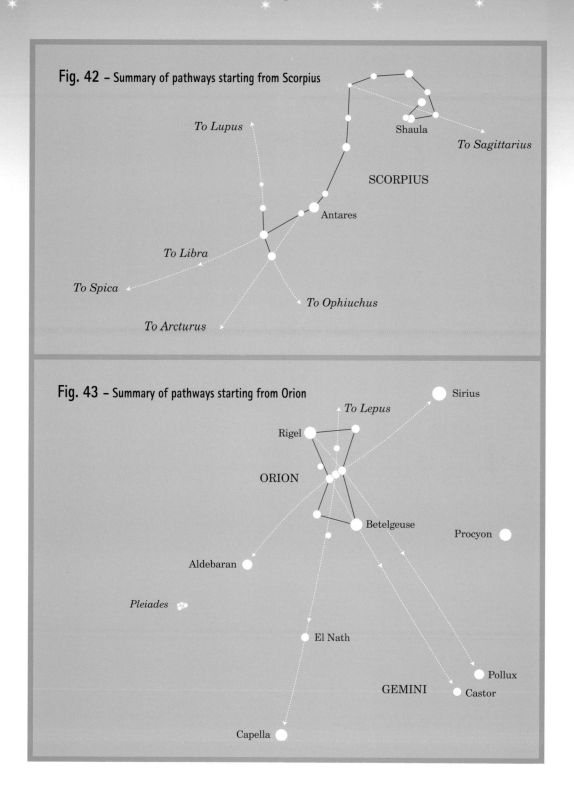

Fig. 42 – Summary of pathways starting from Scorpius

To Lupus

Shaula

To Sagittarius

SCORPIUS

To Libra

Antares

To Spica

To Ophiuchus

To Arcturus

Fig. 43 – Summary of pathways starting from Orion

To Lepus

Sirius

Rigel

ORION

Betelgeuse

Procyon

Aldebaran

Pleiades

El Nath

Pollux

GEMINI

Castor

Capella

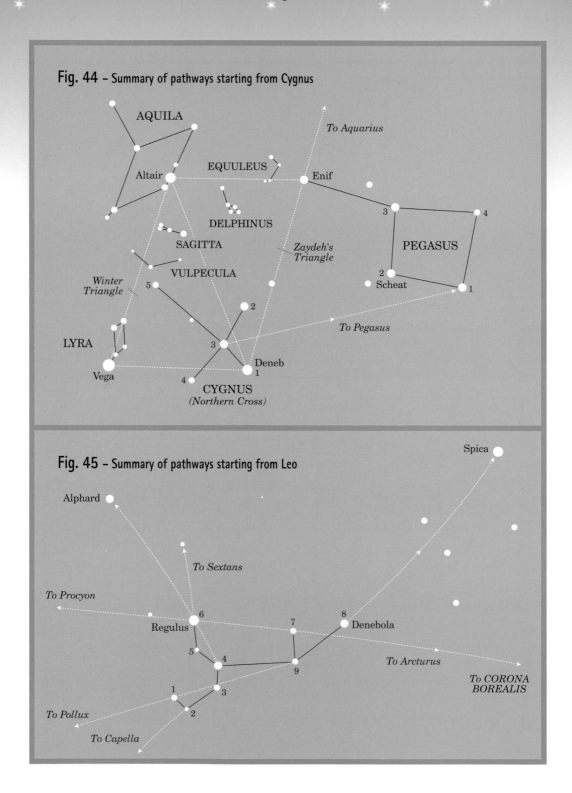

Fig. 44 – Summary of pathways starting from Cygnus

AQUILA

Altair

EQUULEUS

Enif

To Aquarius

DELPHINUS

Zaydeh's Triangle

PEGASUS

3

4

SAGITTA

2

Scheat

1

VULPECULA

Winter Triangle

5

LYRA

3

2

Deneb
1

Vega

4

CYGNUS
(Northern Cross)

To Pegasus

Fig. 45 – Summary of pathways starting from Leo

Spica

Alphard

To Sextans

To Procyon

6

Regulus

7

8

Denebola

5

4

9

To Arcturus

1

3

To CORONA
BOREALIS

To Pollux

2

To Capella

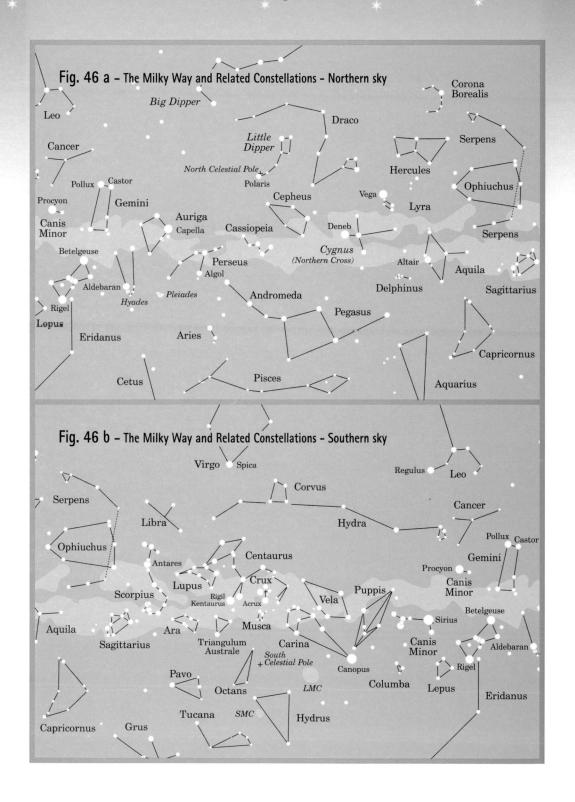

Fig. 46 a – The Milky Way and Related Constellations – Northern sky

Leo
Big Dipper
Draco
Corona Borealis
Cancer
Little Dipper
Serpens
Pollux   Castor
North Celestial Pole
Hercules
Gemini
Polaris
Ophiuchus
Procyon
Cepheus
Vega
Canis Minor
Auriga
Capella
Cassiopeia
Deneb
Lyra
Serpens
Betelgeuse
Cygnus (Northern Cross)
Altair
Aquila
Perseus
Algol
Delphinus
Sagittarius
Aldebaran
Pleiades
Andromeda
Rigel
Hyades
Pegasus
Lepus
Eridanus
Aries
Capricornus
Cetus
Pisces
Aquarius

Fig. 46 b – The Milky Way and Related Constellations – Southern sky

Virgo   Spica
Regulus   Leo
Serpens
Corvus
Cancer
Libra
Hydra
Pollux Castor
Ophiuchus
Centaurus
Gemini
Antares
Crux
Procyon
Scorpius
Lupus
Rigil Kentaurus
Vela
Puppis
Canis Minor
Acrux
Betelgeuse
Aquila
Ara
Musca
Sirius
Sagittarius
Triangulum Australe
Carina
Canis Minor
Aldebaran
Pavo
South Celestial Pole
Canopus
Columba
Rigel
Octans
LMC
Lepus
Eridanus
Tucana   SMC
Hydrus
Capricornus   Grus

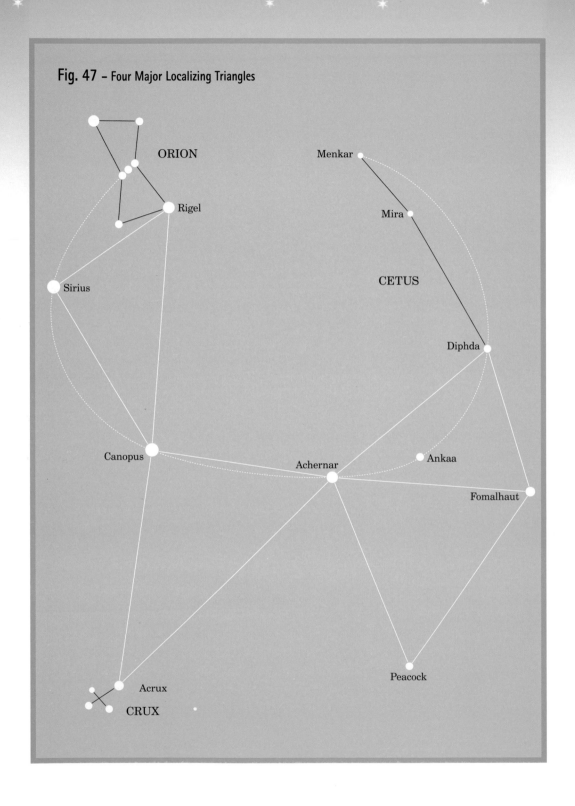

Fig. 47 – Four Major Localizing Triangles

ORION

Rigel

Sirius

Menkar

Mira

CETUS

Diphda

Canopus

Achernar

Ankaa

Fomalhaut

Peacock

Acrux

CRUX

Part 3

Legends of the constellations

The names of many of the constellations are strongly tied to Greek and Roman legends, but the indigenous people living in the Southern Hemisphere did not visualize star groups in the same manner as the ancient Greeks.

The astronomical mythology, except for cosmological myths, of the various Southern Hemisphere people seemed to be primarily concerned with the appearance or disappearance of certain prominent stars, as it may signify the time for a religious ritual or agricultural act. Relatively few elaborate legends have been recorded that relate to a star group, except for myths about the Pleiades, which seemed to have piqued the imagination of many diverse ethnic groups.

Legend of Andromeda

See the legends of Cassiopeia and Perseus.

Legend of Aquarius

Ganymede was a very kind and gentle shepherd boy. One day, while tending his sheep and playing with his dog Argos, the god Zeus sent Aquila, his giant eagle, to swoop down to the plains of Troy to take Ganymede to the temple of the Gods to become Jupiter's favorite water carrier. He was then given ambrosia, the food of the gods, to make him immortal. Wherever Jupiter went Ganymede would accompany him by riding on the back of Aquila the eagle.

Ganymede's kindness was made evident to the gods when he asked Jupiter if he could help the Earth people who were in need of water. Jupiter, who was usually not very kind, was softened by Ganymede's compassion and gave him permission to do as he wished. Ganymede realized that to send a great deal of water to Earth at one time would be dangerous so he decided to send it in smaller amounts and in an intermittent manner in the form of rain. That is how Ganymede, the shepherd boy, became known as the god of rain.

Legend of Aquila

Aquila was Zeus's pet Eagle. Aquila was not only involved with Ganymede in the legend of Aquarius and how Earth was given rain, but was also part of the story of how people got fire.

The Titans were giant gods who were fighting the Greek god of Olympia, Zeus, the new ruler of the world. Prometheus, one of the Titan gods, did not oppose Zeus during the war. After the Titans were beaten he became an advisor to Zeus. While serving Zeus he became aware that the Earth people did not have fire and were not only suffering from the cold but could not have warm food. Zeus was aware of their plight, but was not concerned. Prometheus felt sorry for them and therefore stole a ray of sunshine, hid it in a bamboo container, and sent it down to Earth. With this ray of sun the Earth people made fire to warm their bodies.

Zeus became very angry when he saw that the Earth people were given fire without his permission. He captured Prometheus and chained him to the side of a mountain in the Caucasus where he was to remain forever. He then sent Aquila, his pet eagle, to repeatedly attack Prometheus. After every attack the wound would heal and then would be ripped open again by Aquila. One day when Aquila was about to bite into the abdomen of this kind Titan god, Hercules, who agreed with Prometheus's act of kindness, and who was angered at what Jupiter had done, shot one of his magic arrows into Aquila. Aquila fell seriously wounded. Zeus healed the wounds of his pet eagle and then placed it in the heavens so that it may still soar. It flies near the tail of Cygnus the swan.

Legends of Ara

The constellation of Ara is related to the story of Noah. In the Old Testament it is written that Noah had been told by God that great rains would soon fall until the Earth was covered and all evil would be destroyed. God commanded him to build a giant ark, fill it with all the different seeds, at least two of each form of life and prepare to be cast upon the waters. Noah, his wife and their three sons, Shem, Ham and Japheth and their wives toiled to build the ark as commanded. After all the animals and seeds had been gathered together in the ark the rains came and lasted forty days and nights after which time the Earth was flooded. Even the top of the highest mountain was under water. All life was gone except for those in the floating ark. The floods continued for 150 days at which time a giant wind began to blow and the waters began to subside. Soon their gigantic ark came to rest upon the top of a mountain, possibly Mt Ararat in northern Turkey. The waters continued to fall and after forty days Noah released a raven that flew away only to return and fly in circles over the ark. He then sent out a dove with the realization that if the dove did not return land must be near.

The exhausted dove returned suggesting that most of the land was still covered by the waters. Seven days later Noah released the dove again. After waiting for hours the dove finally returned in the evening with the leaf of an olive tree. Noah waited seven more days, released the dove and when the dove did not return Noah realized that the Earth was once more livable. With great joy they opened the ark and released all the animals. Noah immediately built an altar to pay homage to God. The constellation Ara symbolizes that altar. The dove is symbolized in the constellation Columba. Its brightest star is called Phact, an Arabic word for 'ring dove'.

The flood story is also among the legends of the Sumerian people who lived along the Euphrates in Babylon.

When sailors see the constellation surrounded by clouds they believe it signifies an oncoming gale and stormy seas.

Legends of Aries

Aries was the pet ram of Zeus, the Greek ruler of the heavens. Its coat was made of golden fleece instead of white wool. One day Zeus was looking down on the Earth people when he suddenly noticed that two children on Earth were in danger of being killed. He immediately sent Aries down to Earth to save them. Aries arrived just in time for the children to jump on his back. He then raced to safety. To honor the effort his ram had made, Zeus placed him in the heavens where he can roam freely near the flying horse Pegasus.

Aries also symbolizes the ram caught in the thicket where Abraham was about to sacrifice Isaac. In gratitude to God who sent the angel to stop the hand of Abraham, the ram was placed upon the altar and sacrificed to honor God.

Legend of Auriga

Auriga is portrayed as carrying a goat in his arms as he rides through the sky on his chariot. He is considered the guardian of the shepherds. Shepherds believe that when the constellation of Auriga appears in the sky their sheep will flourish.

It is said that the god Jupiter accidently broke a horn of one of the goats. He apologized for this accident by filling the broken horn with good things. Such a horn has been called the 'Horn of Plenty' or a cornucopia.

Legend of Bootes

Bootes was the son of Demeter, the Greek goddess of agriculture. He was a sensitive young man with a great sense of purpose and social consciousness. When he saw the Earth people struggling to find food he wanted to help them. He realized that if he were to send them food they would always need his help. Instead, he decided to help them to be able to help themselves.

In order for them to be independent, he built the first plough and sent it to Earth. Since then people have been able to plow the land and grow their own food. Because of this great deed for mankind the gods honored him by placing him in the heavens near the handle of the Big Dipper (which is also known as The Plough).

Legends of Canis Major and Minor

Canis Major and Canis Minor were the hunting dogs of Orion. Canis Major was so swift a runner that it could overtake any animal. It was therefore greatly valued by Orion.

The early Egyptians saw the bright star Sirius in Canis Major as the god Anubis, the god with a man's body and the head of a jackal. When Sirius would appear in the sky before dawn, it was the time of the flooding of the Nile, which was of great importance to the farmers who lived along the river since the flood always brought new silt to the land and replenished the soil. It became known as the Dog Star, and the hot days of summer, between July and early September, became known as the Dog Days.

Legends of Canopus in Carina

There are many legends concerning the second brightest star in the sky and the brightest star in the southern hemisphere.

Among the Polynesians Canopus was referred to as Atutuahi, the God of the Heavens, and sung to as the 'Mother of the Moon and the Stars'. It was a major navigational star during their long sea voyages.

Canopus was worshipped by the Egyptians as the god of waters. Several temples were possibly oriented toward Canopus. The desert nomads considered this bright bluish tinted star as a thing of great beauty and called it Al-Suhail, meaning beauty.

Among the Greeks, Canopus was named after the pilot of the great ship of the argonauts that carried Jason, who with the help of fifty argonauts and Medea the sorcerer, successfully captured the golden fleece. Canopus unfortunately died on their return voyage. A monument was erected in his honor around which the city of Canopus was established. The city of Abukir rose upon its ruins. It was in this region that Ptolemy studied the motion of the planets and the stars in the second century CE in Alexandria.

Legend of Cassiopeia

Cassiopeia, the wife of King Cepheus, was the beautiful queen of Ethiopia. She was so proud of her beauty that she became arrogant. She even boasted that her beauty was greater than that of the sea nymphs the Nereids. This boast angered the sea nymphs, who were the daughters of the sea god Nereus, not because they were so vain, but because Cassiopeia failed to appreciate that her external beauty was something she was born with and not something she had achieved. Gratitude for her good fortune would have been acceptable, but not pride. To be proud of something which was not gained through personal effort, but rather with which you were born, showed a poor sense of values.

The Nereids asked the ruling god of the sea, Poseidon (Neptune), to punish Cassiopeia because of her distorted sense of values and her conceit. Poseidon therefore ordered the giant sea monster, Cetus, to destroy their kingdom of Ethiopia.

When King Cepheus and Queen Cassiopeia were informed of Neptune's decision they went to the old wise oracle of Ethiopia for advice. He told them that they must sacrifice their lovely daughter Andromeda to appease the sea gods. Although they were heartbroken, they chained Andromeda to a rock on a cliff overlooking the sea knowing that Cetus, the sea monster, would destroy her.

When Cetus, the sea monster, began to race toward her she screamed for help. Meanwhile, Perseus, who was on the great winged horse Pegasus returning home with the head of Medusa (see the legend of Perseus), heard her cry and immediately flew to her rescue. He arrived just in time to hold up the head of Medusa as Cetus approached. The sea monster was immediately stopped, since anyone who looked directly at the head of Medusa was turned to stone. Perseus carefully placed the head back in its sack, taking care that Andromeda would not look at it. He then unchained Andromeda who fell into his arms. When they gazed into each other's eyes they immediately fell in love.

Although the sea god was angry because the punishment he decreed was not fulfilled, he was so touched by the love of Perseus and Andromeda that he placed them next to each other in the heavens so that their love will always be seen and felt by us on Earth. Poseidon felt that Cassiopeia deserved more punishment. He therefore placed her in the sky in a position where she is condemned to circle around the North Celestial Pole forever, half of the time in an upside-down position.

Legends of Centaurus

Among the people of the Andes mountains in South America those who breed llamas believe good fortune will be theirs if during the birthing season of the llamas (November), the eyes of the celestial llama appears over the southeastern horizon. The eyes of the celestial llama are Hadar and Rigil Kentaurus of Centaurus.

Among the Greeks, centaurs were mythical creatures, half horse half man. Most centaurs were considered to be savage. There was one exception, the son of Kronus and Philyra. When Kronus,

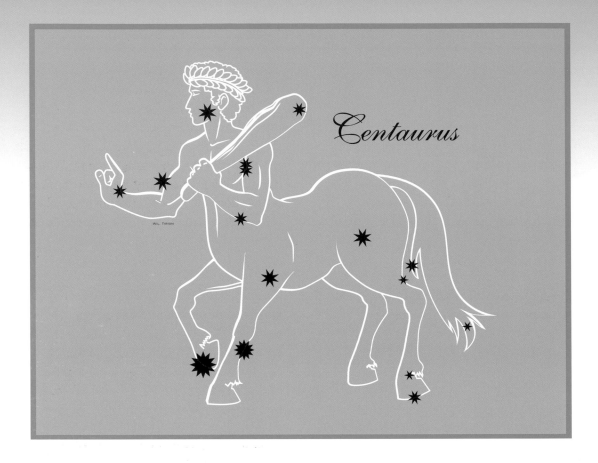

Centaurus

WIL. TIRION

who was married to Rhea, was discovered making love to Philyra, the beautiful sea nymph, he turned himself and Philrya into horses. When Philyra gave birth her child was half horse, half man, a centaur, Philyra was so distressed that she changed her form into a linden tree. Her son was named Chiron. Chiron grew up to be the wisest and gentlest of the centaurs. He lived at the foot of Mt Pelion and established the finest school in Greece. Chiron learned the art of medicine from Artemis the Huntress and the science of astronomy and music from Diana. Jason, Achilles, Hercules, Hylas the Fair, Pollux, Castor and Orpheus all spent part of their childhood studying under Chiron. He was asked to teach Apollo's son Aesculapius the art of healing.

Chiron died as a result of a strange accident. He had taught Hercules how to overcome and capture the wild boar. At the celebration of this deed and during an archery exhibition in which Chiron also participated, a poisoned arrow shot by Hercules accidently struck Chiron in the knee. The pain would not end and was so intense that Hercules begged the god Zeus to grant Chiron's wish to die. After his death Zeus honored him by placing him in the sky where he remains as a symbol of wisdom, goodness and kindness.

Legend of Cetus

Read the legends of Perseus and Cassiopeia.

Legend of Columba (Noah's Dove)

See the Legend of Ara.

Legend of Coma Berenices

Berenice was a beautiful woman whose hair was most glorious. She was married to the Egyptian King Evergetes. When the king left on a dangerous mission Berenice vowed to dedicate her hair to the goddess of beauty if the king were to return unharmed. When he finally returned safely she cut her beautiful hair which Jupiter then placed among the stars. It appears as a cluster of faint stars with a lacy appearance that lies near Arcturus of Bootes and Cor Coroli of Canes Venatici. Coma Berenices has become a symbol of the sacrifices that everyone should be willing to make for their loved ones.

Legend of Corona Australis

Among the desert nomads of ancient Arabia the circle of stars sometimes called the Southern Crown was called the Ostrich Nest. They imagined ostriches in the stars of Sagittarius and Aquila meeting together drinking from the Milky Way, the river in the sky, and then returning to their nest in the circle of stars, the Corona Australis, which lies just south of Sagittarius.

Legend of Corona Borealis

Among the North American Indians this constellation is considered to be a council of chiefs sitting in a semicircle to discuss the future of their people.

In ancient Greece the story is told of Ariadne, the beautiful daughter of King Minos of Crete. When her lover, who was mortal, found out that she had been promised to be wed to a god, he left her on the island of Naxos. Bacchus, the god of vegetation and wine, saw her, fell in love and asked her to marry him. Ariadne did not believe Bacchus was a god. To prove to her that he was a god, Bacchus asked Venus, the Goddess of love, to design a crown of magnificent jewels as a wedding present for Ariadne. When Venus did this for Bacchus, Ariadne was convinced that he was a god and consented to marry him. Bacchus was so ecstatic with joy that he threw the crown of jewels into the heavens where it has been shining ever since.

Legend of Corvus

Corvus was the pet crow of Apollo, the god of sun and music. He was a magnificent bird with a beautiful song. One day Apollo sent Corvus on an errand with instructions to return without delay. Corvus immediately left on his mission. On his return home he saw a fig tree with unripe fruit. This tantalized him, so he waited under the tree for several days until the figs were ripe enough to eat. After eating them he hurried back to Apollo. When asked why it took so long to return, Corvus made up an excuse, but Apollo knew the excuse was false. He was very disappointed that Corvus, whom he had trusted, was not honorable enough to tell him the truth. As punishment Apollo changed the beautiful song of the crow to a hoarse caw sound.

Legend of the Crater

The Crater (the cup) is next to the constellation of Corvus the crow. The Egyptians were very aware of the Crater. They knew that when the Crater rose above the horizon the river Nile, which had been flooding the Egyptian plain, would not rise further and would soon begin to recede.

Legend of Crux

Among the indigenous peoples of Australia is the belief that a giant eagle lives in the sky and that one can see its footprint among the stars of Crux. It nests in the dark patch of sky adjoining Crux now called the Coalsack, and keeps its weapon, the throwing stick, near at hand in the pointer stars of Centaurus, Rigil Kentaurus and Hadar.

There is also the concept that the Coalsack symbolizes evil. This is possibly based upon a legend whereby the Crux formation of stars is seen as a tree. Under this tree an emu would hide in the darkness of the Coalsack area waiting to catch a possum hiding among its branches.

The Zulu people of South Africa considered the Southern Cross as the 'Tree of Life' for it helped those walking in the dangerous bush country at night to find their way.

Legend of Cygnus

Cygnus the swan symbolizes the wonder, the goodness and the dedication which exists in true friendship. Both the constellations of Cygnus and Gemini symbolize the significance of friendship.

Cygnus and Phaethon were very close friends. Phaethon, the son of the mortal woman Clymene, pleaded with his father Helius, the sun god, to help him convince the Earth people that he was the son of a god. Helius agreed to help and therefore told Phaethon that he would grant him any wish. Phaethon immediately asked for permission to drive the four winged horses pulling the chariot carrying the sun. His father pleaded with him not to ask for the almost impossible task of controlling the winged horses, but Phaethon insisted that his father keep his promise. As dawn neared he mounted the chariot with great excitement and began to race across the sky. The great winged horses sensed the inexperienced control of the reins and raced so fast that Phaethon lost complete control of the horses. The chariot swayed so much that the sun was about to fall out of the chariot and burn the Earth. The god Zeus saw what was happening and in order to save the Earth from being destroyed by the heat of the sun, threw a thunderbolt at the chariot. Phaethon lost his balance and fell off the chariot into the roaring river Eridanus. Cygnus saw his friend disappear into the river, and immediately, in spite of the danger, dove into the waters to save him. Helius was so overwhelmed by this act of true friendship toward his son that he changed Cygnus into a diving swan flying along the line of the Milky Way as a symbol of the greatness and importance of friendship.

Legend of Delphinus

Approximately 2600 years ago, on the island of Lesbos, lived a man by the name of Arion. He was a famous poet who was also endowed with a magnificent voice. Arion performed in concerts throughout Greece and Italy. On one of his trips he was sailing to his home in Corinth Greece with all of his valuable prizes. When the sailors realized the value of the prizes on board their ship they became very greedy and decided to steal the wealth and throw Arion overboard. When Arion realized what they were going to do he pleaded with them to allow him to sing one more song while playing his lyre. They agreed. He sang a beautiful song of gratitude to Apollo the god of music and poetry for blessing him with such wonderful talents. Apollo heard the song and knew what was going to happen. He immediately asked the sea god Poseidon to send his messengers, the dolphins, to surround the boat. As Arion sang he noticed the unusual number of dolphins suddenly swarming around the boat. He then jumped overboard with his lyre before the sailors had a chance to grab him. As he was sinking into the deep sea the largest dolphin dived under him, raised him to the surface and then surrounded by the other dolphins they raced away carrying Arion to safety. When Apollo heard about the magnificent action of the dolphin he wished to honor it and therefore placed it in the heavens and placed the lyre of Arion nearby in the constellation of Lyra.

Legend of Draco the Dragon

The ancient Chaldeans, who lived in the region of the Euphrates and Tigris rivers, believed that the dragon Tiamat, the monster of chaos, darkness and evil, lived before the sea and the sky were separated. Tiamat was challenged by the light of the sun and the gods who arose out of the sea of chaos. But he was so powerful and frightening that even the gods gave way. Evil appeared to be winning until Marduk, one of the gods of light, appeared. He had been given all the magical powers that the gods of light and goodness could bring together, and with this power he overcame the dragon, light gained over darkness and good over evil. Tiamat the dragon was placed in the sky as Draco to show all gods and all people that goodness can win.

Legend of Eridanus

The river of Eridanus flows from its origin near Rigel of Orion to flow under Taurus toward Cetus the monster whale. See the legend of Cygnus the swan.

Legends of Gemini

Among the Maori of New Zealand is the tale of twin brothers who were the mortal children of Bora bora. The brothers were extremely devoted to each other and preferred to play with each other rather than with other children. This relative isolation disturbed their parents who then decided to separate the boys. The twins overheard their parent's discussion and decided to run away. They sailed away but their mother followed them. They went from island to island but she was always behind them until they reached Tahiti where they hid in the mountains. She discovered their hiding place and was about to capture them when they climbed to the top of a mountain and flew to the sky where they will always remain close together.

In ancient Greece there is the legend of Castor, a famous horse tamer and soldier, and Pollux, a champion boxer, who were the sons of the Greek god Zeus. They were not only brothers, but very close friends and very adventuresome. At one time they decided to go to sea in order to attack pirates which had been raiding honest seamen. They were so successful in their war against pirates that they became warlike heroes to the sea people, who honored them by carving their images on the prow of their ships. Seamen are aware that during stormy weather sparkling lights may appear on the rigging. When two sparks appear it is an omen that Castor and Pollux are protecting the ship and that the ship will weather the storm. These lights have been called St. Elmo's Fire.

During one of their fights with thieves Castor, who was mortal, was killed. Pollux, who was immortal, was grief stricken and begged Zeus to allow him to be with Castor every other day in the underworld. Zeus was so touched by this request and by Pollux's feeling of true friendship that he not only approved of the request, but placed them in the heavens together so that the Earth people would always be reminded of the preciousness of true friendship.

Legend of Grus

In the Old Testament, Jeremiah (8:7) may have acknowledged Grus as the stork in the heavens. In his condemnation of evil he speaks of the harm one does who does not realize that he or she should go in the direction God has commanded. Jeremiah states that '. . . even the stork knows her appointed season as God has directed'.

Legend of Hercules

Hercules, the son of the god Zeus/Jupiter and of the beautiful mortal woman Alcymene, was the greatest of ancient Greek heroes. He began to show great physical strength as a young child, but more importantly, Hercules revealed a fine sense of character as a young man when he met two women called Pleasure and Virtue. Pleasure promised him enjoyment while Virtue promised him hard work and glory as a doer of great deeds to help mankind. He chose Virtue and was subsequently taught by the wise centaur Chiron.

His deeds included ridding the world of monsters. He fought for 30 days with Leo the Nemean lion before he was able to kill it. He then destroyed the enormous nine-headed water snake of Lernea, which would capture and eat those who ventured near its swamp. The snake was then thrown into the sky and is represented by the constellation Hydra. While fighting the water snake he also destroyed the giant crab, which is now in the sky as Cancer. He captured the wild boar which was destroying the vineyards as well as the fire-breathing bull which was devastating the land.

Hercules continued his work until he had performed twelve deeds. Several years later he was poisoned by mistake with the blood of a centaur. When he died the gods raised him into heaven where he can still be seen as a symbol of one who is dedicated to helping all people by doing good deeds.

Legend of Hydra the Water Snake

Hydra was a nine-headed snake-like monster that killed and ate people as they traveled near the swamps of Lernea. Its blood was poisonous. Hercules was asked to destroy the monster. During their fight every time Hercules would cut off a head another would immediately grow back in its place. He finally succeeded in killing the monster by burning the cut surface on the neck, which prevented regrowth of a head.

Legend of Leo

Leo was a lion who lived on the moon. Food was scarce so he tried to attack one of the horses pulling the chariot of the moon goddess Selene. Leo was thrown off and landed on Earth near Nemea in

Greece where he began to attack people. No person was able to attack this giant beast so they finally called Hercules to destroy this destructive lion. Hercules made a huge knotty club and approached the lion's den. When the lion attacked he swung and struck the lion on the top of its nose. The lion retreated into its cave, but Hercules was so fearless that he followed it into the cave. The roof of the cave was low so that Hercules could not use his club. He therefore jumped on the lion's back and strangled it with his bare hands.

This heroic act was seen by Zeus who honored Hercules by placing the conquered lion in the sky. See the legend of Hercules.

Legend of Libra

Libra symbolizes Astraea, the goddess of justice. She would weigh the souls of men and women on a balancing scale and hold them responsible for their acts. The Sumerians in 200 BCE called it the 'Balance of Heaven'.

Legends of Lyra

Among the Maori the brightest star in Lyra was called Whanui and symbolizes a legend of a love triangle. One night Whanui met the beautiful wife of Rango-Maui. Her name was Pani. Whanui was so overcome by her beauty that although he knew he would be doing wrong he seduced Pani and made love to her. She subsequently gave birth to the sweet potatoes. Her husband Rango-Maui was so disturbed by their presence that Pani allowed him to send the sweet potato children down to Earth. This so angered Whanui that he, in retaliation, sent three kinds of caterpillars down to earth to feed upon the sweet potatoes. As a result, before Whanui, the brightest star in Lyra, appears in the sky at dawn the Earth people store sweet potatoes in the ground to avoid the caterpillars.

Lyra represents the lyre, a harp, which the Greek god Hermes invented and made from a tortoise shell. Its sound was glorious but Hermes was unable to make it sing, so he gave it to Apollo, his brother. Although Apollo was able to make it sing he could not make it sound soulful regardless how much he tried, so he called Orpheus, who was a great musician, to test the harp. When Orpheus picked up the instrument and moved his fingers across the strings the Earth seemed to become silent. All things were listening, the beasts, the birds, the trees and even the flowers turned their faces toward Orpheus. When Apollo saw how music affected all living things he gave the lyre to Orpheus who would play so that people would feel the uplift of music when life seems difficult.

There are times, when the night is very dark and still, that one may look up at Lyra and possibly hear the murmur of Orpheus's song among the sounds of the night.

Legends of Ophiuchus

Among the Babylonians the stars of Ophiuchus and Serpens were thought to portray the sun god Marduk fighting with the dragon Tiamat (Draco). But later in Greek mythology, Ophiuchus was identified with Aesculapius, the Greek god of medicine.

Aesculapius was the son of Apollo and the Thessalian princess Coronis, who died giving birth to Aesculapius. When he was a youth he had such a radiant appearance that everyone knew that he had to be one of the gods. Chiron, the wisest of the Centaurs, taught him the art of medicine. One day Aesculapius observed a snake carrying a herb in its mouth which was used to revive another snake that had been killed. Aesculapius took the herb and with it expanded his knowledge of medicine. Aesculapius was becoming so knowledgeable that the god Zeus feared he would learn how to defeat the death of mortals. To prevent this Zeus felt that Aesculapius must die. He regrettably destroyed him with a bolt of lightning, but then placed him among the stars in the constellation of Ophiuchus. Since then Aesculapius and the snake have been special symbols of healing. Hippocrates, upon whose name all physicians swear an oath to respect the sick, was supposedly a descendant of Aesculapius.

Today's emblem of medicine, the caduceus, commemorates this legend. It is a winged staff with two snakes entwined around it, similar to the staff which was carried by the Greek god Hermes.

Legends of Orion

In the northern tip of Australia in Arnhem Land the Aborigines of Yolngu speak of the time when three famous fishermen, who belonged to the kingfish totem, spent several days at sea trying to catch fish. They were successful, but only in catching kingfish, which they could not eat since it was the totem of their people. They were in a terrible dilemma for their children would go hungry if they did not return with some fish. In desperation they decided to break the taboo against eating kingfish. They resumed fishing and soon caught three more kingfish. The Sun, amazed and angered that they would kill and eat their totem, called upon the clouds, the sea and the wind to create a gigantic waterspout. It was so powerful that it whirled the three fishermen and their canoe high into the sky. To this day they may be seen seated in their canoe as the three stars in a row in Orion. If one looks very carefully, just below the three stars you may be able to see the tiny fish hanging below their canoe.

Among the Ju/Wasi people of Africa is the legend of the god Old/Gao who was hunting for zebras. He finally saw three of them lined up in a row; took aim and shot his arrow, but missed his target. The three zebras escaped and now may be seen as the three center stars of Orion. The arrow may still be seen where it fell. It lies just below the three zebras facing away from them.

In Greek mythology, Hyrieus, a poor farmer who lived in Thebes, was a kind man who frequently befriended strangers although he was poor. One day he helped three unusual strangers. He did not know that they were the gods Zeus, Neptune and Hermes. In return for his kindness he was granted one wish. Hyrieus, who was childless, asked to have a son. The wish was granted and Orion was born. Orion grew up and became a superb hunter for he had been blessed by the gods, but as he became more and more famous as a great hunter he also became insensitive to the animals he hunted. He actually enjoyed the killing of an animal. He did not hunt and kill for necessity. He was so unfeeling about the life of animals that Artemis, the goddess of hunting, sent the giant scorpion (Scorpius) to attack him. He was stung and about to die when Ophiuchus the healer gave Orion an antidote which saved his life. When Orion recovered he realized, after being so close to death, how precious life is and how pitiless and uncaring he had been. He repented and therefore was placed in the heavens with the scorpion whose sting had taught him that all life is precious. See the legend of Sagittarius.

Legend of Pegasus

Pegasus, the magical winged horse of ancient Greek mythology, was the offspring of Poseidon and Medusa. He helped Perseus race through the sky to save beautiful Andromeda (see the legend of Cassiopeia). He was also the steed who rode on the wind carrying the hero Bellerophon through his adventures.

Bellerophon was the son of Corinth and the grandson of Sisyphus. Sisyphus was a selfish and arrogant man who took advantage of people who were less clever. He was therefore punished by the gods and forced to roll a massive stone to the top of a hill, but whenever the top was almost reached the stone would always slip and fall to the bottom. He was forced to continue this struggle for the rest of his life.

Bellerophon was wrongly accused of doing something evil. He was therefore sent on several dangerous missions which he was able to accomplish with the help of Pegasus. As a result he was given the right to keep Pegasus.

As Bellerophon grew older he became arrogant like his grandfather and too proud of his possession of a magical horse that could even ride to the gods. Although Bellerophon was only mortal, he tried to force Pegasus to take him to the top of Mount Olympus so he could mingle with the gods. Pegasus was so astounded at this arrogance that he reared up and threw Bellerophon off his back. Bellerophon fell to Earth while Pegasus flew to the Gods.

Legend of Perseus

King Acrisius of Argos in Greece was told by an oracle that he would someday be killed by his grandson. To prevent this he imprisoned his daughter Danae so that no one could reach her. But Jupiter saw her and fell in love with her. The prison was no barrier. When his daughter gave birth to Perseus, the king put them both in a chest and set them adrift on the sea. The chest did not sink, but eventually landed safely on the island of Seriphus controlled by King Polydectes. When Perseus became a young man he was full of adventure and eager for glory. King Polydectes fell in love with Danae, but realized that her devotion to Perseus would interfere with his courtship of her. He therefore decided to send Perseus away on a long and dangerous mission. He asked Perseus to bring him the head of the Gorgon Medusa. Medusa had once been a beautiful woman, who was so boastful of her beauty that she was turned into a Gorgon, a winged monster with snakes for hair and dragon scales for skin. Whoever looked at Medusa's face would turn to stone.

Perseus needed help to accomplish this task. He coerced three nymphs to help him find the gorgons and to give him the three things he would need to succeed: a pair of winged sandals which allowed him to fly anywhere; a magic helmet which would allow him to see without being seen; and most important he was given a highly polished shield by the Goddess Athena.

After traveling very far he finally found the three Gorgons. They were asleep. He approached them by walking backwards while using his shield as a mirror so he would not look directly at Medusa. He

then cut off her head with a sharp sword given to him by Mercury, and placed it in his bag. Immediately after Perseus had killed Medusa, the winged horse Pegasus arose out of her body.

Perseus jumped on the back of Pegasus and they soared away toward home. As he was passing near Ethiopia he heard the scream of Andromeda who was about to be attacked by Cetus the sea monster. He immediately turned and noticed the beautiful Andromeda chained to a huge rock. Read how he rescued her in the Cassiopeia legend.

Perseus and Andromeda were happily married except for one sorrowful incident. While participating in a discus throwing event, he accidently struck a bystander and killed him. The bystander was his grandfather, King Acrisius, fulfilling the oracle's prophecy that Perseus would cause his grandfather's death. He was so saddened by the tragedy that he gave away the kingdom he inherited.

Legend of the Phoenix

The phoenix is a bird that is the symbol of immortality and the fires within the rays of the sun.

When the time came for the magnificent phoenix to die at the age of 500 years, it arranged a nest for itself among the most fragrant herbs and spices. When the sun rose to its highest point in the sky the phoenix closed its eyes and died. The rays of the sun focused upon the nest and set the nest on fire. When the flames subsided and the ashes cooled a young phoenix emerged and flew toward the setting sun, the source of recurring life.

Legends of the Pleiades

See the legend of Ursa Major.

Among the Aborigines in Central Australia is the legend that the Seven Sisters hungered for some wild figs which could not be found in the sky world, but only on Earth. They therefore came down to Earth. When they arrived they were frightened by the new surroundings and hid in a cave. They were unaware that Nirunja, who lives in Orion, and wanted to make love to the sisters, saw them leave their home in the sky. He secretly followed them down to Earth. When he saw them entering the cave he decided to wait until nightfall when they would be asleep. He then built a camouflage of fig leaves and slowly crept toward the sleeping beauties. As he snuggled among them they awoke and fought their way to the rear of the cave where they escaped through a small crevice in the rock and flew up to their home in the sky. Nirunja, enraged, ran out of the cave, climbed to the top of the mountain and raced after the sisters. Just as he was about to catch them Taurus the Bull, who lives between the seven sisters and Nirunja's home in Orion, awakened from his sleep and faced Nirunja threatening him with his gigantic horns. Nirunja

stared at Taurus, realized he could not get past him, and in frustration returned to his home in the three stars of Orion.

The Maori of New Zealand refer to the Pleiades as Matariki which means 'little eyes', but which also refers to a woman. They visualize its seven visible stars as Matariki and her six daughters. When the Pleiades appears before dawn it is considered the beginning of a new year, at which time the seven women are greeted with songs of hope for the future and songs of tears for the departed. It is a time for festivities and offerings of young shoots of sweet potatoes to Matariki, since she and her daughters watch over and protect their crops.

The Masai of East Africa referred to the Pleiades as the 'rain stars', while the Zulus of South Africa refer to them as the 'digging stars', since they appear at the beginning of the rainy season and denote the time to plow the land.

Among the ancient Greeks the seven stars of the Pleiades were the seven daughters of Atlas and Pleione. It is said that Orion the hunter tried to kidnap Pleione while she was walking with her daughters. Fortunately, they escaped, but as the Pleiades moves across the sky Orion is still in pursuit, never far behind Pleione.

Legend of Sagitta the Arrow

This small constellation, next to the constellation of Aquila, commemorates the

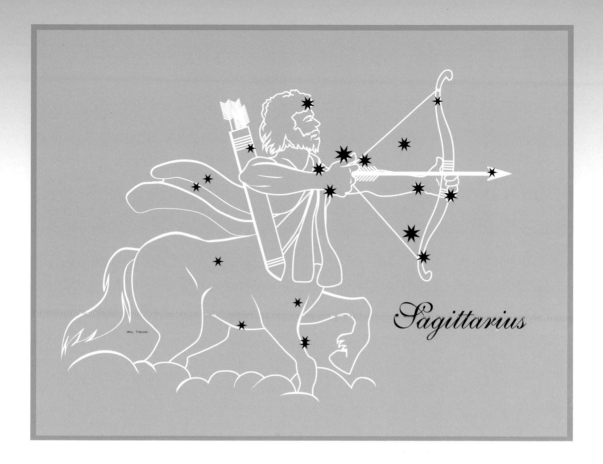

Sagittarius

magic arrow of Hercules which was used to kill Aquila, Jupiter's pet eagle, which was inflicting such agony upon Prometheus. See the legend of Aquila.

Legend of Sagittarius

Sagittarius is considered to be a centaur – half horse, half man, a creature with the power of a horse and the understanding of a person.

Among the ancient Greeks the king of the centaurs was Chiron, the kindest and wisest centaur. He was the teacher of Hercules the great hunter, Aesculapius the father of medicine, Achilles, and Jason who sought the golden fleece. It was Chiron who arranged the stars in the order we now see them.

During one of the travels of Hercules he had the opportunity to befriend Pholos, the son of Chiron, when Pholos was in danger. Chiron was very grateful to Hercules for befriending his son when in need of help. He therefore placed in the constellation Sagittarius a centaur who was a great archer in order to guard and protect Hercules from Scorpius the scorpion.

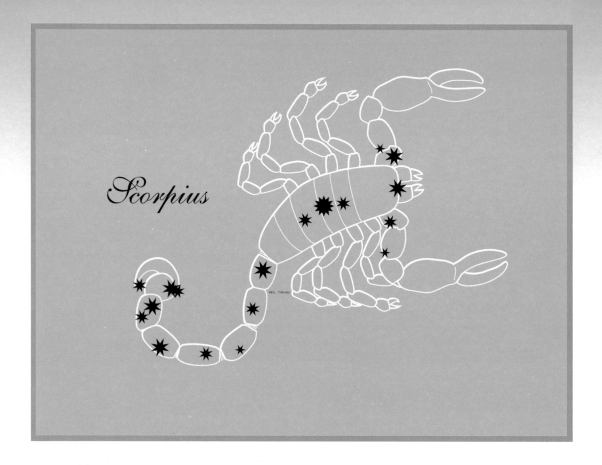

If you look at the heavens at night and watch as Orion sets in the west, you will notice Scorpius rising in the east as if it is following the hunter. But Sagittarius, the archer, follows Scorpius who once attacked Hercules, always ready to attack Scorpius again if it threatens Hercules.

Legend of Scorpius

In China Scorpius was called the 'Azure Dragon'.

The Polynesians of Tahiti tell the story of the boy Pipiri and his sister Rehua. Their mother, who was quite harsh and short tempered, went fishing for the evening meal. It was not until late at night that she caught enough fish for the family dinner. By the time she returned home her husband and children were asleep. They had gone to bed hungry. She awakened her husband and cooked a wonderful meal. He suggested that she awaken the children who had been quite hungry. She refused despite her husband's insistence and put the children's portion away until the following day. Pipiru and Rehua were awakened by the discussion and were severely hurt by their mother's lack of sensitivity to their hunger. They therefore decided to run away from home. Later that night after their parents were asleep they crept out of the house and ran and ran until they reached a tall hill. They climbed the hill and sank to the ground exhausted. They wept because of leaving their home, but they were determined not to return. Their tears flowed until it actually formed a small pool. Meanwhile just before sunrise their mother awakened, saw their bed stained by their tears and immediately awakened her husband to find the children. She saw their tracks barely visible in the early light and followed their small footprints and trail of tears until they reached the top of the hill where the tracks ended. They were looking around, confused for the children were nowhere to be seen until their father looked up and saw Pipiru and Rehua rising to the stars. They decided to follow. When the children saw their parents getting close to them Pipiru asked a giant stag-beetle to help them escape. He placed the children upon his back and with tremendous speed flew up to the stars where they may be seen in the two bright stars in the tail of Scorpius. The giant stag-beetle went to its home in Antares, the bright star in the body of Scorpius. In the South Pacific when parents are unfair or too harsh, children sing a song about Pipiru and Rehua. See the legend of Orion.

Legend of Taurus

The bull was an ancient symbol of worship. He was revered by the Sumerians as the 'Bull of Light', by the Egyptians as Osiris-Apis and was the Golden Calf of biblical times. Taurus, the bull, is the symbol of springtime, which is the time for ploughing and planting, but it is also the symbol of love, which seems to blossom during springtime.

There is a Greek legend that the god Zeus fell in love with the beautiful princess Europa, daughter of King Agenor. Europa had been playing at the seashore. Zeus had been watching her and noticed that when she stopped playing she stood at the edge of the sea and wished that she could go far beyond the horizon. He was so enchanted with her loveliness that he transformed himself into a magnificent white bull. He moved close to the princess and lowered his head. The princess immediately knew that he was offering her the opportunity to fulfill her dream. When she looked into his pleading eyes and felt his wave of love, she climbed onto his back. Zeus, in the form of the bull, dashed into the sea and with great speed swam beyond the horizon to the island of Crete. There he changed himself back to his true form, told her of his love and that he was an immortal god. The princess was so overwhelmed by the intensity and sincerity of his love that she accepted him as her lover. The constellation of Taurus symbolizes this love story.

In the constellation of Taurus is the

star cluster Hyades. Hyades was given a place in heaven because she had nursed Dionysus, the son of Zeus.

The Pleiades, also a star cluster and part of Taurus, were the seven daughters of Pleione and the giant Atlas. There are many legends about the Pleiades. See the legends of the Pleiades and read the legend of Ursa Major, which is also about the Pleiades.

Legend of Ursa Major

When the Earth was very young an American Indian wise man sent his seven sons into the forest to learn how to read the wind. They entered the woods and silently walked while listening to every sound of the wind. When night approached they found a place to rest and to sleep. The stars were bright.

During the night the oldest brother was suddenly awakened by a strange sound. The wind was singing. He could not read the wind song, but as he looked to the stars he saw a bright light flickering in the Pleiades. He was startled for it was flickering in rhythm with the wind song. It appeared to be beckoning to him.

He immediately awakened his brothers to listen to the song and to help read the wind. They joined hands and began to dance. The song became stronger and their dance more intense. Suddenly they began to rise toward the flickering star who was the youngest of the seven sisters of the Pleiades. She had fallen in love with the youngest brother Mizar. Since then, Mizar and his love, given to him by the wind song, can be seen by those with sharp eyes in the handle of the Big Dipper – the home of the seven brothers.

This legend has been derived from a Mongolian and an American Indian legend.

Legend of Virgo

Virgo symbolizes the Earth goddess and the goddess of fertility. It is also the symbol of harvest time.

In ancient Babylonia they spoke of the time when the Earth was dark. The plants would not grow and the animals did not give birth. This was the time when Ishtar, the Chaldean goddess of earth and fertility, went through the seven gates of the underworld to find her husband Tammuz, who had been slain by a wild boar and taken to the underworld. As soon as Ishtar entered the first gate to the underworld, the Earth darkened. When she reached the underworld the Queen of Hell refused to give up Tammuz. When the gods over Earth sent a message to the Queen of the under-world to release Tammuz or be destroyed, Ishtar and Tammuz were sprinkled with magic water, set free and ascended through the seven gates of Hell to Earth. As they reentered Earth spring began, flowers bloomed and the sun warmed the land.

Legends of the Milky Way

The Bushmen of the Kalahari Desert in Central Africa speak of the time when a famous hunter was lost in the dangerous bush. Despite days of searching for his village he failed to find the right path. One night, depressed and fatigued, he rested by the edge of a river and prayed to his gods to help him. Hours later, while looking at the sky, he was suddenly aware of a stream of sparkling stars that seemed to point in one direction. He immediately arose and followed that path in the sky and eventually reached his home. His wife, who realized he must have been lost, was throwing embers of the campfire into the sky to form the path that brought him safely back to her.

The Dogons of Africa had a similar legend in that their god Amma threw pellets of earth into the sky thereby forming the Milky Way.

Among some of the natives of the Andes the Milky Way was considered a river on which the spirits of the dead periodically travelled to return to the land of the living in order to perpetuate communication between the living and the dead.

Among Indigenous Australians there exists a secular description of common knowledge about the Milky Way. A married woman had fallen in love with another man. When she became unfaithful to her husband she tried to hide her affair and therefore lied to him to protect herself and her lover, but he was aware of what she had done. He ordered her to build a large fire. When it was very high he grabbed her and threw her into the flames only to see her immediately float up to the sky where she may be seen as a dark patch in the river of stars, the Milky Way.

The Polynesians also speak of Tane, the son of Rangi who was the sky and light, as the god of the forest, beauty and light as well as the god of the fairies. Rangi and Earth were together in an embrace until Tane separated them, placing the sky high above earth, so that there would be light between them. Tane then threw a basket of stars into the sky to form the Milky Way. Some believe that the Milky Way is the body of Tane's father Rangi.

Some Polynesians consider the Milky Way to be the 'water of life'. They speak of a magnificent blue shark that liked to eat people. It was a pet of the gods. When two young men decided to kill it the gods intervened and placed the shark in the sky where it swims along its river, the Milky Way.

The Milky Way has been thought of as the pathway to the home of Zeus/Jupiter. It was also considered the path of Phaeton's wild ride across the sky in the sun chariot. See the legend of Cygnus.

The Chinese and the Japanese visualized the Milky Way as the silver celestial river. The Norsemen believed the Milky Way to be the path traveled by the departed souls going to Valhalla. In ancient Wales it was the silver road to the castle of the king of fairies, Caer Groyden.

The Algonquin Indians of North America believed it to be the path of the departed spirits on their way to their villages in the sun. The stars of the Milky Way are the campfires that guided them along the path.

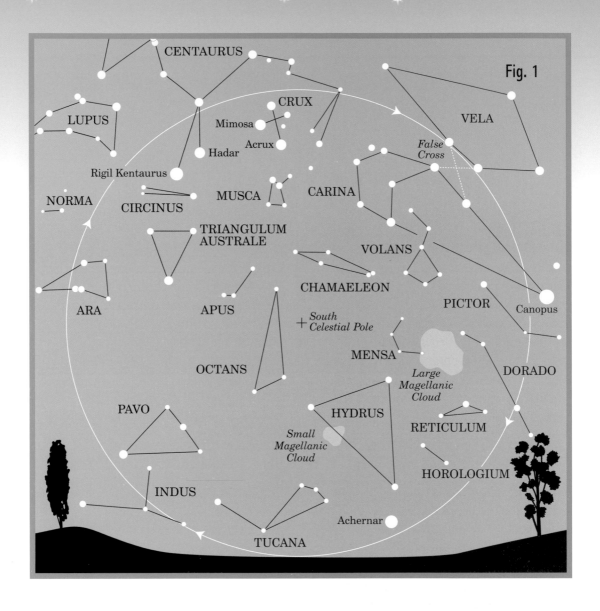

Fig. 1

Part 4

There's more to see!

Circumpolar constellations

If an observer is south of latitude 35 degrees south, certain constellations may always be seen above the southern horizon circling the South Celestial Pole (SCP). The circumpolar constellations include Octans, Chamaeleon, most of Carina, Crux, part of Centaurus, Apus, part of Indus, Musca, Circinus, most of Ara, Pavo, Volans, Tucana, Achernar of Eridanus, Horologium, Reticulum, part of Dorado and Pictor, Mensa, part of Vela, Triangulum Australe, Hydrus, as well as the Large and Small Clouds of Magellan.

The illusion of these constellations rotating around the SCP is due to Earth's rotation around its own axis. If a line were drawn from the south pole through the center of the earth through the north pole and extended to the celestial sphere it would end approximately 0.8 degrees from Polaris in the north and 13 degrees from Star 2 of the constellation Hydrus in the south.

Test of vision

Although there are 200 billion stars like our Sun in our galaxy, we can only see about 2500 of them at one time above the horizon under ideal conditions.

Color test

Stars vary in color. These colors depend on a star's temperature (just as the color of a flame depends on its temperature). The coolest stars are red, the hottest blue. If you look closely, you can see some of these colors for the brightest stars. Try looking at some of those listed below.

Reddish
Antares in Scorpius
Aldebaran in Taurus
Belelgeuse in Orion

Yellowish
Capella in Auriga
Rigil in Centaurus
Canopus in Carina

White
Sirius in Canis Major
Fomalhaut in Piscis
  Austrinus
Altair in Aquila

Blue White
Vega in Lyra
Rigel in Orion
Regulus in Leo
Spica in Virgo
Castor in Gemini
Hadar in Centaurus
Acrux in Southern Cross

Orange
Arcturus in Bootes

Yellow White
Procyon in Canis Minor

Star brightness test

Most people can see fifth magnitude stars, some can see sixth and even seventh magnitude stars. Remember, the higher the magnitude, the dimmer the star. Orion is an excellent testing ground. Look for the hazy nebula (a mass of dust and gas) above Orion's belt (marked with an arrow on Fig. 2). The numbers show the magnitude (brightness) of each star.

How many stars can you see in the Pleiades? Five is good, six is very good and seven is excellent. To study the faint stars of Centaurus will help to solidify one's mental image of the figure.

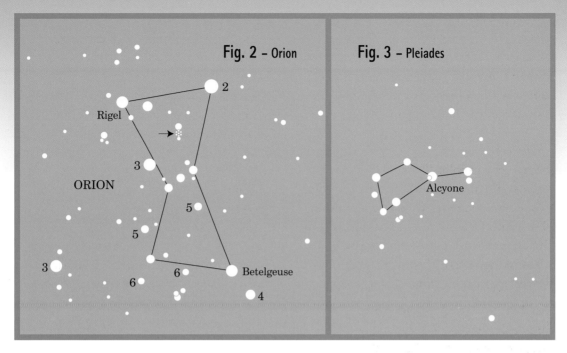

Fig. 2 – Orion

ORION
Rigel
2
3
5
5
3
6
6
Betelgeuse
4

Fig. 3 – Pleiades

Alcyone

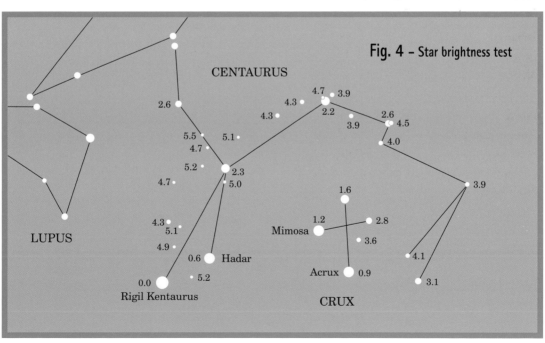

Fig. 4 – Star brightness test

CENTAURUS

2.6
4.3
4.3
4.7
3.9
2.2
3.9
2.6
4.5
4.0
5.5
5.1
4.7
5.2
2.3
5.0
3.9
4.7
1.6
4.3
5.1
1.2
2.8
4.9
Mimosa
3.6
0.6
Hadar
4.1
LUPUS
Acrux
0.9
3.1
0.0
5.2
Rigil Kentaurus
CRUX

The twinkling of the stars is not the star itself twinkling, but the effect of atmospheric air currents that break up the rays of light. If the air is very turbulent even the planets may twinkle.

clouds of dust and gas that may appear as a dark area against a background of stars, or with a faint glow if a luminous star is nearby. Galaxies are star systems similar to our own Milky Way. They are sometimes referred to as 'Island Universes'.

Binocular sights

Binoculars are rated by their magnifying power and by the diameter of the objective lens (front lens). The ability of binoculars to increase the amount of light seen by the eye is dependent upon the diameter of the objective lens. The maximum opening of the pupil of the human eye is approximately 8 millimeters. This permits you to see stars with a magnitude of six and possibly seven. A binocular with a rating of 7×32 means that its light-gathering ability, based upon a 32 millimeter objective, would be approximately four times (32/8) as great as your naked eye. This greater light-gathering power enables one to see star clusters, nebulae, galaxies and double stars that you could not see with your naked eye, and single stars with a magnitude of nine or ten.

A star cluster is a group of stars that appear very close together. A double star is actually two stars so close together that they appear as one star. Nebulae are

Location of planets

As the planets orbit the Sun they all lie on approximately the same plane as the orbit of Earth except for Pluto. That plane, or orbital path, is called the ecliptic plane. As the Earth orbits the Sun, thirteen constellations are sequentially hidden from view by the Sun. These constellations comprise the Zodiac. Astrologers prefer to consider only twelve constellations as comprising the Zodiac.

Some general points of interest.

1 There are five planets visible to the naked eye: Mercury, Venus, Mars, Jupiter and Saturn.

2 Mars has a distinctive reddish tint.

3 Mercury is difficult to see since it is always seen just after the Sun has set and therefore appears against a fairly bright sky background.

4 The best time to see Venus or Mercury is after sunset in the west or before dawn in the east.

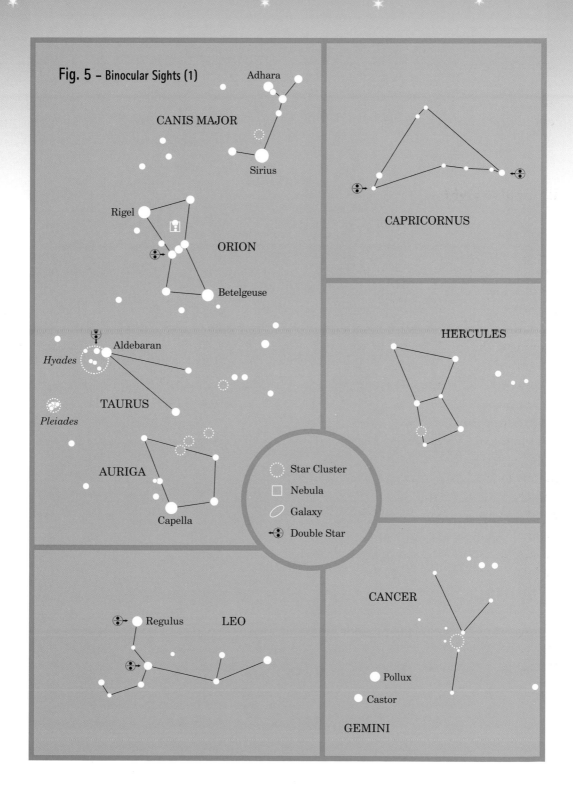

Fig. 5 – Binocular Sights (1)

Adhara

CANIS MAJOR

Sirius

Rigel

ORION

Betelgeuse

Aldebaran

Hyades

TAURUS

Pleiades

AURIGA

Capella

CAPRICORNUS

HERCULES

Star Cluster
Nebula
Galaxy
Double Star

LEO

Regulus

CANCER

Pollux

Castor

GEMINI

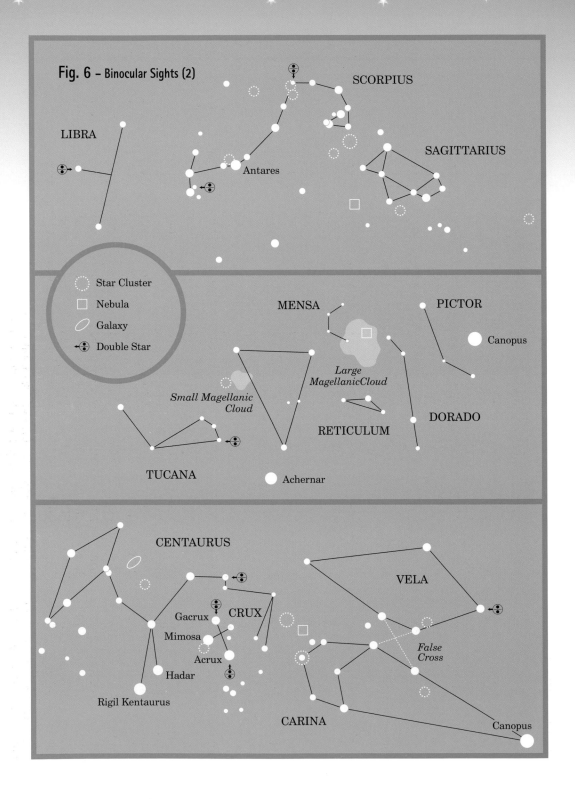

Fig. 6 – Binocular Sights (2)

LIBRA

SCORPIUS

SAGITTARIUS

Antares

Star Cluster

Nebula

Galaxy

Double Star

MENSA

PICTOR

Canopus

Large Magellanic Cloud

Small Magellanic Cloud

RETICULUM

DORADO

TUCANA

Achernar

CENTAURUS

VELA

Gacrux

CRUX

Mimosa

False Cross

Acrux

Hadar

Rigil Kentaurus

CARINA

Canopus

Planet locations

2000	Jan.	Feb.	Mar.	Apr.	May	Jun.	Jul.	Aug.	Sep.	Oct.	Nov.	Dec.
Venus	LIB	SGR	CAP	AQR	ARI	TAU	GEM	LEO	VIR	LIB	OPH	SGR
Mars	AQR	AQR	PSC	ARI	TAU	TAU	GEM	CNC	LEO	LEO	VIR	VIR
Jupiter	PSC	PSC	ARI	ARI	ARI	TAU	TAU	TAU	TAU	TAU	TAU	TAU
Saturn	ARI	ARI	ARI	ARI	ARI	TAU	TAU	TAU	TAU	TAU	TAU	TAU

2001	Jan.	Feb.	Mar.	Apr.	May	Jun.	Jul.	Aug.	Sep.	Oct.	Nov.	Dec.
Venus	AQR	PSC	PSC	PSC	PSC	PSC	TAU	GEM	CNC	LEO	VIR	LIB
Mars	VIR	LIB	SCO	OPH	SGR	SGR	OPH	OPH	SGR	SGR	CAP	CAP
Jupiter	TAU	TAU	TAU	TAU	TAU	TAU	TAU	GEM	GEM	GEM	GEM	GEM
Saturn	TAU	TAU	TAU	TAU	TAU	TAU	TAU	TAU	TAU	TAU	TAU	TAU

2002	Jan.	Feb.	Mar.	Apr.	May	Jun.	Jul.	Aug.	Sep.	Oct.	Nov.	Dec.
Venus	SGR	CAP	AQR	ARI	TAU	GEM	LEO	VIR	VIR	LIB	VIR	VIR
Mars	AQR	PSC	ARI	ARI	TAU	GEM	GEM	CNC	LEO	LEO	VIR	VIR
Jupiter	GEM	GEM	GEM	GEM	GEM	GEM	GEM	GEM	CNC	CNC	CNC	LEO
Saturn	TAU	TAU	TAU	TAU	TAU	TAU	TAU	TAU	TAU	TAU	TAU	TAU

2003	Jan.	Feb.	Mar.	Apr.	May	Jun.	Jul.	Aug.	Sep.	Oct.	Nov.	Dec.
Venus	LIB	SGR	SGR	AQR	PSC	ARI	TAU	CNC	LEO	VIR	LIB	SGR
Mars	LIB	OPH	SGR	SGR	CAP	CAP	AQR	AQR	AQR	AQR	AQR	AQR
Jupiter	CNC	CNC	CNC	CNC	CNC	CNC	LEO	LEO	LEO	LEO	LEO	LEO
Saturn	TAU	TAU	TAU	TAU	TAU	GEM	GEM	GEM	GEM	GEM	GEM	GEM

2004	Jan.	Feb.	Mar.	Apr.	May	Jun.	Jul.	Aug.	Sep.	Oct.	Nov.	Dec.
Venus	CAP	AQR	PSC	TAU	TAU	TAU	TAU	TAU	GEM	LEO	VIR	LIB
Mars	PSC	ARI	ARI	TAU	TAU	GEM	CNC	LEO	LEO	VIR	VIR	LIB
Jupiter	LEO	LEO	LEO	LEO	LEO	LEO	LEO	LEO	VIR	VIR	VIR	VIR
Saturn	GEM	GEM	GEM	GEM	GEM	GEM	GEM	GEM	GEM	GEM	GEM	GEM

2005	Jan.	Feb.	Mar.	Apr.	May	Jun.	Jul.	Aug.	Sep.	Oct.	Nov.	Dec.
Venus	OPH	SGR	AQR	PSC	ARI	TAU	CNC	LEO	VIR	LIB	SGR	SGR
Mars	SCO	SGR	SGR	CAP	AQR	AQR	PSC	PSC	ARI	TAU	ARI	ARI
Jupiter	VIR	VIR	VIR	VIR	VIR	VIR	VIR	VIR	VIR	VIR	VIR	LIB
Saturn	GEM	GEM	GEM	GEM	GEM	GEM	CNC	CNC	CNC	CNC	CNC	CNC

Positions are for the first day of each month

OPH = Ophiuchus, SGR = Sagittarius, CAP = Capricornus, AQR = Aquarius, PSC = Pisces, ARI = Aries,
TAU = Taurus, GEM = Gemini, CNC = Cancer, LEO = Leo, VIR = Virgo, LIB = Libra, SCO = Scorpius.

Navigational stars

Stars have guided navigators as they crossed the oceans and explored foreign lands. Fifty-seven stars have been designated as navigational stars

Star	Constellation
Acamar	Eridanus
Achernar	Eridanus
Acrux	Crux
Adhara	Canis Major
Aldebaran	Taurus
Alioth	Ursa Major
Alkaid	Ursa Major
Al Na´ir	Grus
Alnilam	Orion
Alphard	Hydra
Alphecca	Corona Borealis
Alpheratz	Andromeda-Pegasus
Altair	Aquila
Ankaa	Phoenix
Antares	Scorpius
Arcturus	Bootes
Atria	Triangulum Australe
Avior	Carina
Bellatrix	Orion
Betelgeuse	Orion
Canopus	Carina
Capella	Auriga
Deneb	Cygnus
Denebola	Leo
Diphda	Cetus
Dubhe	Ursa Major
El Nath	Taurus
Eltanin	Draco
Enif	Pegasus
Fomalhaut	Piscis Austrinis
Gacrux	Crux
Gianah	Corvus
Hadar	Centaurus
Hamal	Aries
Kaus Australis	Sagittarius
Kochab	Ursa Minor
Markab	Pegasus
Menkar	Cetus
Menkent	Centaurus
Miaplacidus	Carina
Mirfak	Perseus

Star	Constellation
Nunki	Sagittarius
Peacock	Pavo
Pollux	Gemini
Procyon	Canis Minor
Rasalhague	Ophiuchus
Regulus	Leo
Rigel	Orion
Rigil Kentaurus	Centaurus
Sabik	Ophiuchus
Schedar	Cassiopeia
Shaula	Scorpius
Sirius	Canis Major
Spica	Virgo
Suhail	Vela
Vega	Lyra
Zubenelgenubi	Libra

Minor constellations

Constellations dimmer than 3.5 magnitude

Antlia	<4.3
Apus	<3.8
Caelum	<4.5
Camelopardalis	<4.3
Cancer	<3.9
Coma Berenices	<4.3
Corona Australis	<4.1
Crater	<3.6
Equuleus	<3.9
Fornax	<4.1
Horologium	<3.9
Lacerta	<3.8
Leo Minor	<3.8
Mensa	<5
Microscopium	<4.7
Monocerus	<3.9
Norma	<4
Octans	<3.8
Pisces	<3.6
Pyxis	<3.7
Sculptor	<4.3
Scutum	<3.9
Sextans	<4.5
Volans	<3.8
Vulpecula	<4.4

Constellations index

Constellations marked with an asterisk contain one or more stars that are at least as bright as magnitude 3.